Pharmaceutical Chemical Analysis: Methods for Identification and Limit Tests

Pharmaceutical Chemical Analysis: Methods for Identification and Limit Tests

Ole Pedersen

Taylor & Francis
Taylor & Francis Group
Boca Raton   London   New York

A CRC title, part of the Taylor & Francis imprint, a member of the
Taylor & Francis Group, the academic division of T&F Informa plc.

Published in 2006 by
CRC Press
Taylor & Francis Group
6000 Broken Sound Parkway NW, Suite 300
Boca Raton, FL 33487-2742

International Standard Book Number-10: 0-8493-1978-1 (Hardcover)
International Standard Book Number-13: 978-0-8493-1978-5 (Hardcover)
Library of Congress Card Number 2005052909

Library of Congress Cataloging-in-Publication Data

Pederson, Ole.
    Pharmaceutical chemical analysis : methods for identification and limit tests / Ole Pederson.
        p. ; cm.
    Includes bibliographical references and index.
    ISBN 0-8493-1978-1
    1. Drugs--Analysis--Laboratory manuals. 2. Drugs--Standards--Laboratory manuals. 3. Pharmaceutical chemistry. I. Title.
    [DNLM: 1. Pharmaceutical Preparations--analysis--Laboratory Manuals. 2. Pharmaceutical Preparations--standards--Laboratory Manuals. 3. Drug Compounding--standards--Laboratory Manuals. 4. Quality Control--Laboratory Manuals. 5. Reference Standards--Laboratory Manuals. QV 25 P371p 2005]

RS189.P39 2005
615'.1901--dc22   2005052909

Taylor & Francis Group
is the Academic Division of Informa plc.

**Visit the Taylor & Francis Web site at
http://www.taylorandfrancis.com**

**and the CRC Press Web site at
http://www.crcpress.com**

Preface

The obvious potential severity of wrongfully giving an already weak patient the wrong medication, or giving medicine with a harmful content of unwanted substances, led relatively early (compared to other industries) to formal governmental demands on the quality of pharmaceutical products and pharmaceutical ingredients. One aspect of these demands includes chemical tests specified in pharmacopeias defining requirements on identity and chemical purity with which raw materials bound for pharmaceutical dosage from production must comply. The methodology of the tests adapted has in principle reflected the current knowledge and technological expertise of the pharmaceutical society at the time of adaptation of the test methods in question. Early monographs relied largely on appearance, taste, and smell as well as microscopic characterizing, whereas recently adapted monographs are often based on sophisticated and expensive analytical techniques such as capillary electrophoresis, X-ray spectroscopy, and an extensive use of high-performance liquid chromatography.

There is, however, some tardiness in the process of replacing methods relying on outdated analytical techniques with more modern ones for several reasons. First, the pharmaceutical industry is in general a rather conservative one with respect to bringing in new technology, again for several reasons. Huge economical interests are at stake so no one wants to jeopardize the product development time line by using technology that has not already proven its quality. So, newer methods are often run in parallel with older ones for a long time, thereby not fully releasing its cost saving and knowledge enrichment potential. In addition, the pharmaceutical industry has increasingly been the subject of goods manufacturing procedure (GMP) requirements, and are therefore, the subject of periodic inspections. Also, one must justify methodology used when filing for approval of a new drug product. Obviously it will be easier to prove the validity of a well known and widely used technique and a manufacturer could be tempted not to spend time and effort in justifying the capability of novel techniques to ensure consistent and sufficient drug product quality. Also, it is a substantial bureaucratic maneuver to effect the change in a *European Pharmacopoeia* general test referendum referenced in many individual monographs in a supernational organization like the European Pharmacopoeia Commission. A last reason for not replacing old methods is that some tests actually perform quite well and one could question why something simple and economical should be replaced with expensive instrumental techniques.

This, however, leaves us with a number of rather old chemical methods, which are based on analytical principles, like colorimetry, using selective reagents and precipitation-based techniques, which have largely been abandoned in modern analytical chemistry. The basic assumptions and theoretic background of these techniques are to only a limited extent a part of the training of the contemporary and very instrumentally orientated analytical chemist. Even less likely will he or she have enough knowledge about specific color reagents. The literature dealing with the techniques is so old that nothing is found in an electronic literature search and information is therefore less accessible. The specific knowledge collected by national medicinal agencies is to a large extent closed to the public.

The past few decades have seen a massive increase in GMP requirements, and this has been reflected in recruitment especially in the analytical quality control laboratory. Very often, within the department, job candidates have been screened for their GMP experience and attitude as compared to analytical chemistry skills. The prerequisites of the typical quality control analytical chemist are to understand the tests performed and thereby being capable of solving any occurring problems, which very often does not equal the outside pressure and potential economic implications an unsolved analytical problem might cause. This book is intended to aid the quality control chemist, but also in general to make knowledge about these tests easily accessible to academics and technicians working with these tests. It is organized into two parts dealing with the tests of *European Pharmacopoeia* 2.3.1 and 2.4, identification and limit tests, respectively. Each part contains three general chapters where subjects relevant to all the pharmacopoeial identification/limit tests are discussed, and then the individual tests are described and discussed in sections organized and named as in the *European Pharmacopoeia*. The test sections start with a few short remarks about the purpose and rationale of the tests, followed by a review of the physical and chemical characters of the ion or compound that is the target of the test. Then the chemical background and logic of the individual procedural steps of the test are described, with formulas and reaction, including remarks about the strengths and weaknesses in terms of specificity, ruggedness, and potential procedural pitfalls. All chapters include a list of references that could be the starting point of further investigations.

The methods of the two *European Pharmacopoeia* general test chapters are among the oldest in the pharmacopoeia. Many of the tests on the simple inorganic ions are based on selective precipitation using only simple inorganic reagents. Many of these tests were developed in response to the needs of the mining industry in the nineteenth century. A wealth of information on the specificity, strengths, and weaknesses of these techniques can be found in older inorganic qualitative and quantitative analytical chemistry literature, but very often it is not possible to find evaluations on methods carried out exactly as adapted by the pharmacopoeia. The use of colored derivatives began in the early 20th century, gained much attention with the development of more sophisticated optical equipment during the middle of the century, until their application declined dramatically toward the end of the century as a result of the success of chromatographic techniques. Such methods are dealt with in analytical chemistry handbooks and also in scientific journals published prior to the 1980s.

Acknowledgment

The author wishes to thank
Tina, QC GEA 2002–2004, and Albert Bentz.

About the Author

Ole Pedersen is a pharmacist of the Danish University of Pharmaceutical Sciences and has throughout his career worked with various aspects of analytical chemistry in the pharmaceutical industry. His work led to his broad expertise in the fields of analytical quality control and analytical development, and a thorough knowledge of GMP and regulatory affairs requirements.

Table of Contents

Part I

Identifications

chapter one

Precipitation in identifications

In qualitative tests based on selective precipitation, the identity of the substance in question is verified by ensuring that it gives a precipitate under the defined conditions. In many cases it is a more or less explicit demand of the test that the precipitate have a defined appearance, such as for example a dense gelatinous or crystalline appearance. Whether a substance gives a precipitate in the conditions the tests specify depends on its solubility under these exact conditions and also on the kinetics of the given precipitation. The appearance of a precipitate, apart from the color, is largely determined by its particle size distribution and by its affinity toward the water molecules and ions of the surrounding solution.

Precipitate selectivity

A substance precipitates from a given solution when its solubility is exceeded. In the case of ionic substances, this means that the factor Q, which is the product of the molar concentrations of the salt forming ions, is higher than its solubility product. The solubility product is a constant defined as the product of the molar concentrations of the individual salt forming ions, in a saturated solution of the salt in question. The solubility product formula for the general salt A_aB_b is

$$Q = [A^{z+}]^a \times [B^{z-}]^b$$

$$Ksp = [A^{z+}]^a \times [B^{z-}]^b$$

In many cases, the negative logarithm to the solubility product is used to make the vast span of solubility products encountered easier to manage.

$$pKsp = -\log (Ksp)$$

The important thing to have in mind is that precipitation is facilitated whenever the solubility product is exceeded, also in the cases where the ratios of individual ion concentrations are very different from the ones found in a saturated solution of the salt in question. This means, when referring to the general formula, that precipitation will take place even in cases of a very low A concentration compared to the ones seen in a saturated solution, if only B concentration is high enough. This is interesting especially in salts where protons (H^+) or hydroxide (OH^-) participates, since test solution pH then influences Q directly. A deviation in test solution pH of 1 or 2 units perhaps does not seem very much in a simple test such as an identification, but the pH scale being logarithmic means that [H^+] or [OH^-] is actually 10 or 100 times from the desired value. Depending on the stoichiometry of the salt in question, this has a large effect on Q.

It also means that if ions that participate in a given salt are added to the solution from unintended sources, this will lower the solubility of the salt by raising the test solution ion product. This is the so-called common ion effect.

It should also be noticed that the stoichiometric coefficients end up as exponents in the formula for the solubility product. The ion product Q is therefore more sensitive toward changes in concentrations of ions present in the salt formula in a stoichiometric ratio higher than 1.

Finally, it should be remembered that the solubility product in the form presented above is a simplification of the real formula, which uses ion activity a instead of ion concentration:

$$Ksp = a_A{}^a \times a_B{}^b$$

These are obtained by multiplying the molar concentration by activity coefficients. The activity coefficients are symbolized by γ:

$$Ksp = [A^{z+}]^a \gamma_A{}^a \times [B^{z-}]^b \gamma_B{}^b$$

The value of γ is given by the Debye-Huckel equation that incorporates, among other parameters, both the ionic strength of the solution and the valence of the ion. It is a number between 0 and 1, where 1 applies for very dilute solutions. The effect of a high ionic strength in a solution is that γ declines and deviates from 1. When multiplied by the concentration it gives a lower activity and ultimately a lower Q than would have been the case if molar concentration had been used, so that the concentration needed to reach the solubility product is higher.

This means that salts will display a higher solubility in a solution of high ionic strength than predicted by experiments in weaker ionic solutions or pure water. All the dissolved ionic compounds contribute to the solution's ionic strength, especially the ones of a higher valence than 1. So the unintended presence of ionic substances brought there as sample preparation residues or by simple contamination might prevent a precipitation or disturb the basis of a selective precipitation.

The kinetics of a precipitation is dependent on the actual salt in question and on the degree of supersaturation in the solution. This means that even though the ion product might not have been lowered by ionic strength to an extent that precipitation is eliminated, it might have lowered the degree of supersaturation sufficiently to slow down the onset of precipitation. It again might alter the appearance of the precipitate.

The exact way a selective precipitation is performed has a strong influence on the selectivity obtained. One could, for example, identify a cation forming amphoteric hydroxide by first precipitating it with sodium hydroxide and then redissolving it in excess sodium hydroxide. In principle, this procedure can be designed in two different ways. One could slowly add sodium hydroxide until precipitation occurs and then further add sodium hydroxide until the precipitate redissolves. Or one could first add a defined amount of sodium hydroxide that knowingly should precipitate the actual cation and then further add a defined amount of sodium hydroxide that knowingly should redissolve the actual cation in question.

In terms of judging the ruggedness and selectivity of the test, these two principles are quite different. In the first case a high degree of ruggedness is achieved. Even if some contaminants in the test solution, or the counter-ion in the substance to be examined, delay precipitation, eventually, it will take place as further sodium hydroxide is added. This situation could entirely prevent precipitation from occurring as a consequence of adding the defined amount of sodium hydroxide dictated in the second case, thereby giving a false positive reaction. The selectivity, on the contrary, is more limited in the first procedure. All cation-forming amphorteric cations will give a positive reaction, regardless of the amount of sodium hydroxide needed to precipitate and redissolve the various cations.

In general, when discriminating inorganic ions by their solubility as different salts and in different chemical environments, one should recognize that seemingly small variations in test conditions can change the selectivity and robustness of the test. Variations in sample concentration, pH, and ionic strength of the test solution and presence of a counterion can alter the situation from what is intended.

Precipitate appearance

Precipitates have often been placed in one of three different categories based on their macroscopic appearance. A precipitate can be colloidal, crystalline, or curdy gelatinous. The kind of precipitate given by a specific salt has no correlation to its degree of insolubility but is determined by the particle size distribution of the precipitate, and its affinity toward the water molecules and ions of the surrounding solution.

Identification tests are most often designed to give fast results. In general this means that test solution and reagent concentrations are adjusted so that a relatively high degree of supersaturation of the desired salt is achieved. Under such conditions, chances are that the precipitation will occur immediately, but it also means that the precipitate formed will consist of a high number of very small particles. Often particles are so small that a colloidal suspension is achieved. Colloids are particles with diameters in the range of 1 μm down to 0.001 μm. They are so small that they form suspensions so stable that they are sometimes called colloidal solutions. Particles of a size below the interval tend to form true solutions, and particles of a size above the interval tend to form less stable real suspensions.

Colloids are drawn together by van der Waals forces, but these are balanced by the repulsive power of an electrical double layer that builds up on their surfaces. The ratio of these forces largely influences the appearances of such suspensions. In the core of an ionic particle, cations and anions are well ordered in stochiometric units. The salt forming anion or cation, whichever is in excess in the solution, will be absorbed on the surface of the colloid. This anion or cation, called the primary absorbed ion, will give the colloids a positive or negative charge, causing opposites to repel each other. This hinders aggregation of the individual particles into larger structures. But the electrical charge also attracts ions of the opposite charge to the immediate vicinity of the particle surface, building up an electrical double layer. If this ion, called the counter-ion, is loosely attracted, giving a diffuse double layer, its presence will not neutralize the charge of the primary absorbed ion, creating repulsion between the particles. But if for some reason, the counter-ion is very closely attracted to the primary absorbed ion layer, the result is a narrower double layer, causing the particle charge to be neutralized. Such particles have a tendency to aggregate into a more assembled precipitate, which appears curdy if the salt in question has a limited affinity toward water and gelatinous if it has a high affinity toward water.

If, for example, chloride is precipitated with an excess of silver nitrate, colloidal silver chloride is formed having silver as the primary absorbed ion and nitrate as the counter-ion. This is often represented by the formulas below. The number of dots between the primary absorbed ion and the counter-ion symbolizes the proximity of the two.

$$AgCl{:}Ag^+ .. NO_3^-$$

or

$$AgCl{:}Ag^+ ... NO_3^-$$

The process of reducing the distance between the primary absorbed ion and the counter-ion, and thereby settling the colloidal precipitate is called flocculation or coagulation. This process can be achieved by heating the colloidal suspension or by raising its ionic strength by adding an inert ion. The process of heating is sometimes referred to as digestion and is also used with crystalline precipitates, but then the aim is to facilitate crystal growth and not coagulation. However, since the aim in both cases is to change a small particle size precipitate into a more settled one that is easier to handle, names are used somewhat interchangeably.

The fact that flocculation is the result of high ionic strength is of some importance in relation to identification tests, since ionic strength is influenced by sample preparation steps, the nature of the counter-ion of the substance to be examined, and the test solution contaminants. The concentration that is necessary for a given ion to cause flocculation depends on the nature of the ion, especially its valance. The higher the valance, the stronger its flocculation ability.

Some salts have a tendency to form a precipitate consisting of fewer but larger particles, even though precipitation is initiated from a highly supersaturated solution. Or, they may form a colloidal precipitate, which rapidly changes in particle size distribution. Such a precipitate will appear crystalline, and particles will, unlike in a colloidal suspension, quickly settle. The factors determining the particle size distribution of a precipitate are discussed in some detail in Chapter 4 "Precipitation in Limit Tests."

It is of some importance for the operator performing qualitative tests to have a basic understanding of the various types of precipitates, since a certain appearance of the obtained precipitate is sometime an explicit requirement of the test. Even if no specific type of precipitate is obligatory, some insight into their differences can prove valuable in the case of analytical problems in performing a test.

One obvious way of investigating a suspected analytical problem is to perform a positive control, using a test on a substance of known identity. If a known sample of the substance to be examined is at hand, it would be the obvious choice. In cases where this is not obtainable, other compounds containing the substance to be examined can be used, as for example another salt of the wanted cation. It should then be remembered, however, that precipitate appearance can deviate from what one would have obtained using the right substance, owing to effects given by the other part of the molecule. One example is a precipitate formed as directed in 3.32 Sulfates might deviate when made on the sulfate salt of a large organic cation rather than on sodium sulfate. Another example is that differences in ionic strength in test solutions of different substance cause varying degrees of coagulation of otherwise similar precipitates.

To introduce the appearance of a colloidal, curdy, gelatinous, or crystalline to an inexperienced operator, one could use a few precipitates of established nature. Performing the first part of test (a) in 3.16. Chlorides on sodium chloride, without shaking the test tube, will yield a colloidal precipitate, which is easily coagulated to a curdy precipitate, for example by shaking, heating, or adding an inert ion. Performing test (a) for 3.32. Sulfates on sodium sulfate will yield a colloidal precipitate that can be digested by heating to a crystalline precipitate. A good example of a gelatinous precipitate is achieved by performing the first part of the test 3.35. Zinc on zinc chloride, since the zinc hydroxide is gelatinous, as are many other metal hydroxides.

chapter two

Color reactions
in identifications

About half of the identification tests are based on a methodology in which a selective chemical reaction performed on the substance to be examined yields a colored entity or even a colored precipitate that is directly recognizable as positive. These reactions are most commonly referred to as colorimetric reactions, although the name is also incorrectly used for photometric and spectrophotometric determinations. Historically, in a photometric analysis, an unknown sample is compared to a series of standards of known concentrations using a natural light photometer, whereas spectrophotometry uses a UV-visible spectrophotometer with monochromatic light. In colorimetric determinations, on the contrary, samples and standards are compared and judged visually.

Since the identification tests are carried out without the use of positive controls or standards, they owe their robustness and reliability to the selectivity of the chemical reaction and the recognizability of the color produced. Especially the selectivity of the chemical reaction will naturally depend entirely on the actual reaction in question, but a few general remarks should be made.

In principle, there are several ways of using a color reaction in an identification test.

- The substance to be examined is changed to a colored compound by the action of a reagent that does not become a part of the new entity formed. One example is test (b) for 3.19. Iodides, where identification is brought about by the violet or violet-red color produced when iodide is oxidized by dichromate to iodine.
- The substance to be examined changes an uncolored reagent into a colored compound without becoming a part of the new molecule formed.
- The substance to be examined reacts with a reagent forming a colored compound in which both the reagent and the substance to be examined

are included. These could be covalently bound condensation products or more loosely associated chalets and salts. Examples of these are numerous: the condensation of certain amines and the reagent β-napthol in 3.5. Amines, primary aromatic, or the cobalt complex formed in 3.10. Barbiturates non-nitrogen substituted.

In some cases, the actual reaction will include a combination of the above.

An ideal chemical derivative color reaction will possess a number of qualities. It should preferably be unidirectional (not reversible), highly selective, and sensitive. The reaction product should be stable and have an appearance that is very different from the appearance of the substance to be examined and the reagent itself. Finally, it should be rugged toward small variations in the determination procedure. When dealing with an identification reaction even performed without comparing the result obtained with a standard, obviously the selectivity is of the greatest importance. Here it is important to remember that color reactions in their nature are at the best selective and not specific.

In the case of a condensation between the substance to be examined and a reagent molecule, a functional group of the reagent reacts with a functional group of the substance to be examined. Since no functional group is found in one substance exclusively, the reagent will in principle react with all sample molecules having this particular functional group in its structure. Some degree of selectivity can be achieved, however, if the substance to be examined reacts very easily and under very mild conditions compared to other compounds of its class due to, for example, a steric hindrance aspect. Other reactions, on the contrary, are not even truly selective with respect to functional groups but will under the right conditions give a positive reaction with compounds of other classes.

One example is the Legal reaction used for identification of 3.21. Lactates and 3.17. Citrates. The reagent used here, sodium nitroprusside, can react with all ketones and aldehydes having an unsaturated α-carbon. However, the reaction is more or less likely to occur depending on the actual compound's ability to shift into its enol resonans form and also depending on steric hindrance in the actual compound. Another example is the test for esters, where the ester bond is hydrolyzed and the carboxylic acid converted into a hydroxamic acid, which gives a distinct colored complex with iron(III). This reaction has been used for the determination of compounds characterized by a number of functional groups including many carboxylic acids, acid anhydrides, acid halides, acid amides, and imides.

It is therefore evident that color reactions cannot afford a selective, not to mention specific, identification of a compound. Their value lies in the relative ease with which they can be performed, and in the fact that they are unconstrained by expensive analytical equipment. Usually they are used together with two or more other identifications, which combined, provide a reasonable certainty about the identity of the substance to be examined.

chapter three

Tests

Chapter 2.3 of the *European Pharmacopoeia* general Chapter 2.3 includes identifications for the monograph 2.3.1, and also identifies reactions of ions and functional groups. These identifications consist of 35 simple tests each aiming at identifying, with some certainty, that the examined substance contains the specific ion or functional group giving name to the test. The tests share the common feature that they can be performed at the laboratory bench top without using analytical equipment. The tests are divided into two categories based upon the analytical/chemical principles (precipitations and color reactions) by which they were founded. However, they are very different in terms of selectivity, sensitivity, and ruggedness toward interferences. The aim of this chapter is to give the reader some insight into the principles of the individual tests and to be the starting point for further investigations. This starting point is based either upon determining whether one such study is needed as part of an out of specification investigation or whether a decision to reference the test when developing a new monograph is necessary.

The typical monograph references one or more of the tests of 2.3.1. Identification reactions of ions and functional groups normally include three or more identifications. When combined, these identifications should result in a satisfactory level of confidence without being 100% discriminatory; as for example, a NMR spectroscopic elucidation. The monograph covering concentrated, hydrochloric acid includes two identification tests: a test for chloride and a pH measurement test, using indicator paper. A positive test merely reveals that the examined substance contains chloride and is strongly acidic. Obviously, even combined, these two tests will not afford an absolute certainty that the substance is really hydrochloric acid. But within the concept of a GMP compliant production, where there are also high quality demands the on raw material supplier, it is judged to suffice. It is within this set of thinking that the selectivity, sensitivity, and ruggedness of the individual test should be judged.

3.1 Acetates (European Pharmacopoeia 2.3.1)

The method is primarily used for the identification of acetate entities where these are present as the salt and not as a chemically bound acetyl group. This constitutes the difference from the complementary test 3.2. Acetyl, which identifies the substance to be examined as a compound containing an acetyl group that upon acidic hydrolysis yields acetic acid. An exception is the use of 2.3.1. Acetates in the monograph of the polymer polyvinyl acetate, where the acetic acid group has been attached to an alcohol group by esterification. In such a case, the group is chemically bound to the substance to be examined and has to be liberated by hydrolysis prior to analysis. The salts typically tested vary from the simple sodium salt to acetate salts of complex organic molecules such as hydroxocobalamin acetate. Reference to acetates is made in 10 monographs, and test (a) and test (b) are referenced almost equally.

Acetic acid forms soluble salts with most cations, exceptions being the salts of iron, aluminum, and chromium, which are insoluble. The salts of silver and mercury (I) are sparingly soluble. Acetic acid is a weak carboxylic acid with a pKa of 4.75 (25°C) and this means that a 0.1 M solution is about 3% dissociated.

Test (a)

The substance to be examined is heated with an equal quantity of *oxalic acid R*. Acid vapors with the characteristic odor of acetic acid are liberated, showing an acid reaction (2.2.4).

If the salt of a weak acid is treated with a stronger acid, it is liberated from its salt and forced onto the acid form. Since oxalic acid is a stronger acid with a pKa of 1.25, acetate will shift acetate to free acetic acid, which is volatile and has a characteristic smell. Oxalic acid is not volatile. The water necessary for the reaction comes from the crystal water of the oxalic acid reagent. The reaction is exemplified with sodium acetate (Figure 3.1.1) as shown.

Since the substance to be examined in this test is identified by its smell, great care should be exercised so that the test is not performed on a substance that could give off gases that would be hazardous to inhale. When elaborating new monographs, the inclusion of organoleptic analysis is discouraged.[1]

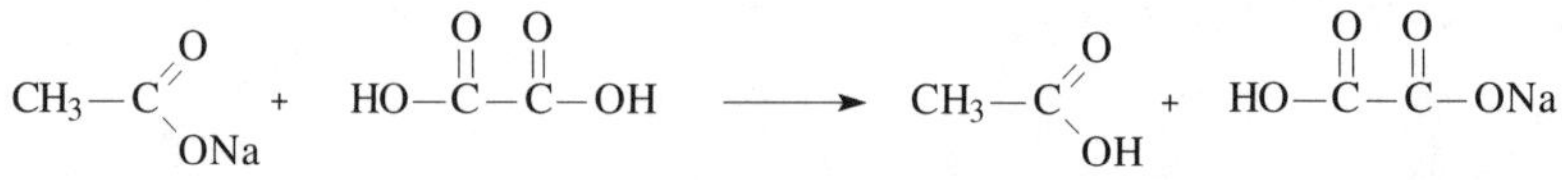

Figure 3.1.1 Liberation of acetic acid by oxalic acid.

Test (b)

The test, which was first described by Damuor in 1857, is based on the blue complex that is formed by iodine with the basic precipitate of lanthanum acetate.[2]

The procedure is simple: successively add to the test solution 0.25 ml of *lanthanum nitrate solution R*, 0.1 ml of *0.05 M iodine*, and 0.05 ml of dilute *ammonia R2*. When the mixture is heated until boiling, the formation of a blue precipitate or a blue coloration of the solution indicates a positive reaction.

If lanthanum nitrate and ammonia are added to the test solution without iodine being present, a white precipitate of basic lanthanum nitrate develops. The precipitation initiates slowly at room temperature but is speeded up by heating. This precipitate will, in the presence of acetate ions, form a complex with iodine that can have the form of a blue solution or a blue precipitate.

Not much is known about the mechanism of the reaction or the nature of the complex. Since iodine forms blue complexes with a large number of colloidal solutions of substances, including starch, one theory is that the physical colloidal structure of the lanthanum precipitate is responsible for the ability to form the complex. Another theory is that a specific chemical structure of the analyt (acetate) is responsible. The discussion is not conclusive, but the investigations performed to reveal the nature of the test disclose that the procedure is quite specific and robust.

The precipitation must be carried out in a pH interval of 9–11 to facilitate precipitation and avoid dispropionation of the iodine to iodide and iodinate, which allegedly begins to take place at pH values above 11. If oxidants are present iodine will be changed into iodide. The presence of anions with which lanthanum (III) forms insoluble salts will often disturb the test; BO_3^{3-}, $C_2O_4^{2-}$, PO_4^{3-}, SO_4^{2-}, and F^- could be mentioned. Also ions that form complexes with lanthanum, e.g., citric acid, will destroy the result. NO_3^-, Cl^-, Br^-, and I^- do not interfere even if present in a 30 to 40 times excess to acetate, but they will weaken the intensity of the color.[3,4]

The test is found to be very robust toward large changes in the described proportions of the reagents. It is very insensitive toward the amount of iodine added, but eventually at very high concentration the natural yellow color of iodine will affect the blue color. A very high concentration of KI (added to the iodine for solubilization) can change the color from blue to very dark blue, or even brown or yellow. It should be noted, however, that a very large deviation from normal reagent concentration is needed. The speed of precipitation is linked to acetate concentration and ionic strength of the test solution. Vigorous boiling of the precipitate solution can weaken the intensity of the coloration due to loss of iodine.[5]

Only propionate and monofluoro acetate are mentioned to react as acetate thereby giving a false positive result.[6] Other small carboxylic acids have been found to change the color of the complex, propionate, i-butyrate, n-butyrate, n-valerinate, iodo acetate, monochloro acetate, dichloro acetate, trichloro acetate, tribromo acetate, and p-toluol sulfonate.[7] The test is very sensitive and has been used to reveal the presence of even very small amounts of acetate.

References

1. Council of Europe Pharmacopoeia Technical Guide, Council of Europe, 1990.
2. Damour, M.A., Note sur le sous-acétate de lanthane iodé, *Compt. Rend. Acad. Sci.*, 43, 976, 1857.
3. Hartke, K., 2.3.1. Identitätsreaktionen (10.Lfg.1999), in *Arzneibuch-Kommentar zum Europäische Arzneibuch*, Band I, *Allgemeiner Teil*, Hartke, K. et al., Wissenschaftliche Verlagsgellschaft, Stuttgart/Govi-Verlag — Pharmazeutischer Verlag, Escbor.
4. Feigl, F., *Tüpfelanalyse*, Band 2, *Organisher Teil*, Vierte deutsche auflage, Akademishe Verlagsgesellschaft, Frankfurt am Main, 1960, p. 344.
5. Krüger, D. and Tschirch, E., *Die Blaufärbung des "Basischen Lanthanacetats" mit Jod. Eine Hochempfindliche Reaktion auf Acetat-Ion*, Berichte d.D. Chem. Gessellschaft, 62, 2776, 1929.
6. Vejdělek, Z.J. and Kakáč, B., *Farbreaktionen in der Spectrophotometrischen Analyse organischer Verbindungen*, Band II, VEB Gustav Fischer Verlag, Jena, 1973, p. 489.
7. Krüger, D. and Tschirch, E., *Die Gefärbten Jodverbindungen Basischer Salze Seltener Erden. Ein Beitrag zum Jod-Stärke-Preoblem (II. Mitteil.)*, Ber., 63, 826, 1930.

3.2   Acetyl (European Pharmacopoeia 2.3.1)

The test identifies the substance to be examined as a compound containing an acetyl group that upon acidic hydrolysis yields acetic acid. The method bases its chemistry on the same reaction as acetates method (b). Reference to acetyl is made in 12 monographs, of which acetate esters of steroids constitute a large group. In a few monographs it is specified that the material should be treated as a *substance hydrolyzable only with difficulty*. Several classes of compounds are capable of releasing acetic acid through hydrolysis, but the most obvious examples are esters and amides. The hydrolysis of an ester and an amide is shown below using cortisone acetate (Figure 3.2.1) and N-acetylthryptophan (Figure 3.2.2) as an example.

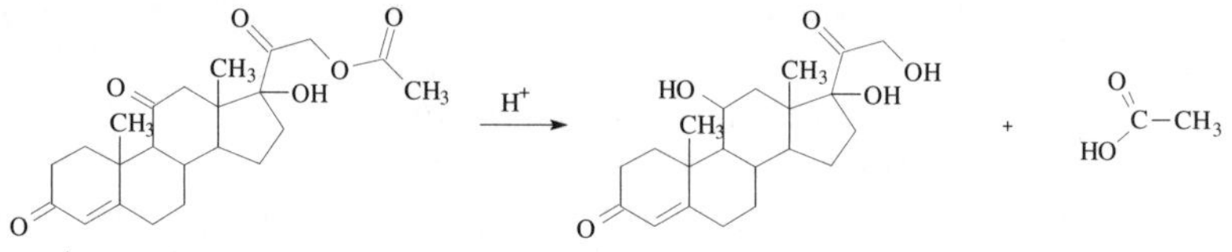

Figure 3.2.1 Cortisone acetate hydrolysis.

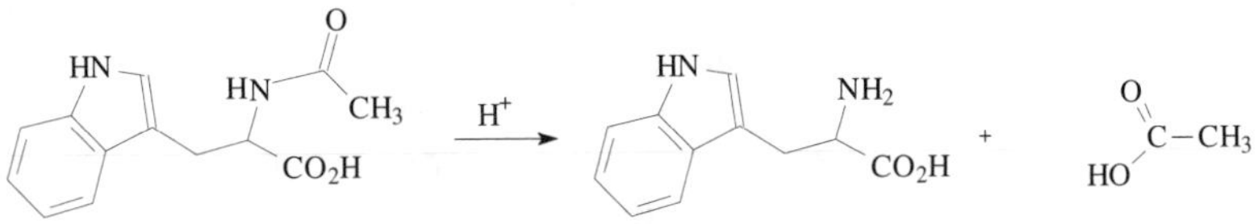

Figure 3.2.2 N-acetylthryptophan hydrolysis.

In a test tube about 180 mm long and 18 mm in external diameter, about 15 mg of the substance to be examined, or the prescribed quantity, is placed and 0.15 ml of *phosphoric acid R* is added. The tube is closed with a stopper through which passes a small test tube about 100 mm long and 10 mm in external diameter containing *water R* to act as a condenser. On the outside of the smaller tube, hang a drop of *lanthanum nitrate solution R*. Except for substances hydrolyzable only with difficulty, place the apparatus in a water-bath for 5 min and then take out the smaller tube. The drop is removed and mixed with 0.05 ml of *0.01 M iodine* on a tile. At the edge, 0.05 ml of dilute *ammonia R2* is added. After 1 to 2 min, a blue color develops at the junction of the two drops; the color intensifies and persists for a short time. For substances hydrolyzable only with difficulty, heat the mixture slowly to boiling over an open flame and then proceed as prescribed above.

When the substance to be examined is dissolved in phosphoric acid in the larger test tube (Figure 3.2.3) and heat is applied, acetic acid is released and evaporated. Some of the gas will dissolve in the drop of lanthanum

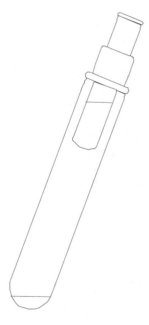

Figure 3.2.3 Test tube assembly.

nitrate solution hanging below the smaller test tube. When iodine is added and the drop is brought in contact with a drop of dilute ammonia, a blue complex or precipitate is formed as described in 3.1. Acetates.

It is worth mentioning that substances containing salts of acetic acid or acetic acid itself will also give a positive reaction in the test. Otherwise, the comments given in 3.1. Acetates on specificity and interference apply here as well. It should be recognized, though, that the chance of interference must be considered even lower than in the before-mentioned, since the test solution containing the substance to be examined does not come in contact with the reagent.

3.3   Alkaloids (European Pharmacopoeia 2.3.1)

The test identifies the substance to be examined as an alkaloid. Alkaloids are a very large class of substances with a somewhat vague definition. Traditionally alkaloids have been identified with the use of four analytical reagents:

- Mayer's reagent, a potassiomercuric iodide solution
- Wagner's reagent, a solution of iodine in potassium iodide
- Hager's reagent, a picric acid solution
- Dragendorff's reagent, a potassium bismuth iodide solution

One definition of an alkaloid is that it is a substance that gives a positive reaction to the above reagents. Another definition is given by Trease and Evans: "Typical alkaloids are derived from plant sources, they are basic, they contain one or more nitrogen atoms (usually in a heterocyclic ring) and they usually have a marked physiological action on man or animal."[1]

From this definition, a few general properties can be expected. Owing to the nitrogen atom they are basic (alkaloids means alkalilike), and therefore they have a higher water solubility at low pH than at high pH. Besides alkaloids constitute a large group with a variety of properties. Usually alkaloids are distinguished through a classification based on the amino acid from which they are derived, for example the ornithine alkaloids. The chemical diversity of the alkaloid group can be illustrated by the structures of morphine (Figure 3.3.1) and hyoscyamine (Figure 3.3.2) given below. Both of them are substances used in therapy. At the present about 10 monographs make reference to alkaloids. The identification test of the European *Pharmacopoeia* makes use of the Dragendorff's reagent, which is a solution of potassium iodobismuthate, $KBiI_4$.

A test solution is prepared by dissolving a few milligrams of the substance to be examined, or the prescribed quantity, in 5 ml of *water R*. When dissolved, *dilute hydrochloric acid R* is added until an acid reaction occurs (2.2.4). Then 1 ml of *potassium iodobismuthate solution R* is added, an orange or orange-red precipitate is formed immediately.

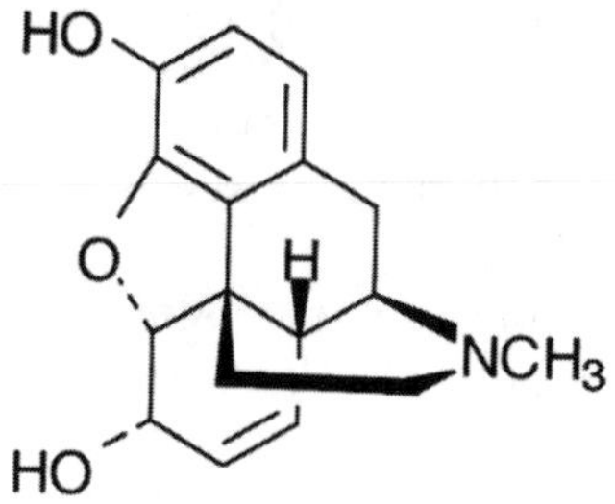

Figure 3.3.1 Morpine.

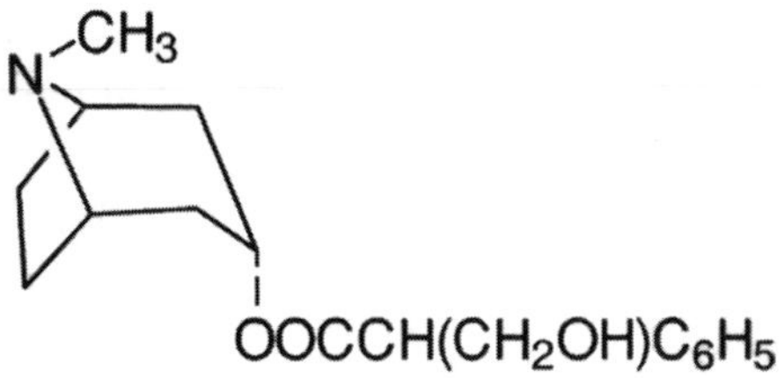

Figure 3.3.2 Hydroscyamine.

The acid reaction occurs, as defined in the test, when methyl red changes from yellow to orange or red at about 5.1 pH or when bromothymol blue changes from blue to yellow at about 7.1 pH. This means that most tertiary and quaternary amines are found in the ionic acid form, and this allows the thereby cationic alkaloid to precipitate with anionic iodobismuthate. The salt is believed to be constituted not of monomeric entities, of for example one nitrogen base and one iodobismuthate, but of oligomeric cations and anions.

$$(\text{alkaloid}^+)_m \ (\text{BiI}_4{}^-)_m$$

A positive reaction is obtained with many tertiary and quaternary amines, but not with most primary and secondary amines. All the alkaloids described in monographs making reference to alkaloids are tertiary and quaternary.

Obviously, alkaloids are not the only group of substances containing tertiary and quaternary amine groups, so not much specificity can be claimed. Most of the literature dealing with the specificity of the Dragendorff,s reagent concentrates on the risk of false positive reactions by other secondary metabolites potentially present in plant material. Here cumarines, hydroxyflavones, some triterpenes, and cardenolides could be mentioned.[2] Even proteins can sometimes give positive reactions.

When judging the specificity of the reagent and its usefulness in relations to analysis on pure active pharmaceutical ingredients, one should keep in mind that the original purpose of the reagent was somewhat different. It was used, together with other reagents, to screen whether a plant might contain something that could prove to be an alkaloid.

References

1. Evans, W.C., *Trease and Evans' Pharmacognosy,* 13th ed., Bailloère Tindall, London, 1989, p. 544.
2. Hänsel, R., Sticher, O., and Steinegger, E., *Pharmakognosie — Phytopharmazie,* 6. auflage, Springer-Verlag, Berlin, 1999, p. 955.

3.4   Aluminum (European Pharmacopoeia 2.3.1)

This method identifies the presence of aluminum ions, Al^{3+}, in the substance to be examined. Reference to aluminum is currently made in seven monographs, all of them being inorganic salts. Only a few of them describe substances that are soluble in water; in the remaining monographs the remaining Al^{3+} has to be released with an acidic solvent. Aluminum's only ionic form is the aluminum(III) ion, Al^{3+}. Its halides, nitrates, and sulfates are water soluble. Solutions of these salts are mildly acidic because of hydrolysis of the hydrated ion $[Al(H_2O)_6]^{3+}$, which is the form of Al^{3+} in watery solutions.

$$[Al(H_2O)_6]^{3+} + H_2O \rightarrow [Al(H_2O)_5OH]^{2+} + H_3O^+$$

A larger number of cations, especially the ones placed in the periodic table around the border between the metals and the nonmetals, form amphoteric hydroxide. That is, they form insoluble hydroxides in alkaline environments, which dissolve in excess sodium hydroxide, and this fact is used in the test. To enhance the selectivity of the identification, aluminum's reaction toward sulfide and ammonium ions is tested as well.

In the last three steps of the analysis, we test whether the cation present forms insoluble hydroxides upon the addition of sodium hydroxide, which dissolves in excess, and that the reaction is reversed by ammonium chloride. Since a fairly large number of cations (Be^{2+}, Pb^{2+}, Sn^{2+}, Sn^{4+}, and Sb^{3+}) have the same characteristics, a false positive reaction by one of these is initially eliminated by precipitation with S^{2-}.

In this first step of the analysis, the 2 ml test solution is added to 0.2 ml dilute *hydrochloric acid R* and about 0.5 ml *thioacetamide reagent R*. In the acidic environment, thioacetamide will liberate sulfide ions by hydrolyze. The thioacetamide reagent is preferred to SH_2 gas and Na_2S solutions as a source of S^{2-} ions, since it is safer to handle.

Aluminum's sulfide salt shows a high degree of solubility, so no precipitation occurs. But at the low pH of the test solution, where the concentration of the sulfate ion is very limited, cations with very small solubility products, e.g., lead sulfide, will precipitate.

$$Pb^{2+} + S^{2-} \rightarrow PbS_2\downarrow$$

In this way, false positive reactions by Pb^{2+}, Sn^{2+}, Sn^{4+}, and Sb^{3+} should be eliminated.

In the second step, dilute *sodium hydroxide solution R* is added dropwise. This leads to the formation of a white gelatinous precipitate of aluminum hydroxide.

$$Al^{3+} + 3OH^- \rightarrow Al(OH)_3\downarrow$$

In case any of the above mentioned and hopefully eliminated ions did not precipitate with S^{2-} at low pH, they will do so now at high pH where the concentration of S^{2-} is much greater. But the precipitates most likely will be very different in appearance from the white gelatinous aluminum hydroxide, because of their color and colloidal characteristics.

Owing to aluminum's ability to form amphoteric hydroxides, the aluminum hydroxide precipitate dissolves on the further addition of dilute *sodium hydroxide solution R* as a hydroxo complex is created.

$$Al(OH)_3\downarrow + OH^- \rightarrow [Al(OH)_4]^-$$

Owing to the initial S^{2-} precipitation at low and high pH, only Al^{3+} and Be^{2+} could be the cause of the observations. The precipitate of aluminum hydroxide can by aging transform into a form that is insoluble in excess sodium hydroxide, so the further addition has to be performed shortly after precipitation.

In the last step the precipitate of aluminum hydroxide is reformed by the addition of *ammonium chloride solution R*. This is a result of the downward shift in pH caused by the acid reagent.

$$[Al(OH)_4]^- + NH_4^+ \rightarrow Al(OH)_3\downarrow$$

This step eliminates a false positive reaction caused by the presence of zinc, which would stay in solution because of formation of the soluble complex $[Zn(NH_3)_4]^{2-}$ (see 3.35. Zinc). The step is, however, most likely redundant, since Zn^{2+} should already have been eliminated in the sulfide precipitation. But since Be^{2+} does not form a soluble ammonium complex but reprecipitates a beryllium hydroxide, it could cause a false positive. Realistically, though, one must say that the presence of a beryllium salt in a raw material for pharmaceutical production is not very likely.

A false positive reaction due to interfering ions seems unlikely, since four assumptions have to be fulfilled. If this is suspected, a test performed on known substances, e.g., aluminum chloride or a working standard of the substance to be examined, as the exact appearance of the precipitates formed will give valuable information. One cause of false negative reactions could be contamination with reagents, which prevents the manipulation of the test solution's pH. If sample preparation has failed and interfering anions are present in the test solution, this of course would give problems as well.[1-3]

References

1. Vogel, A.I. and Svehla, G., *Vogel's Qualitative Inorganic Analysis*, 6th ed., Longman Scientific and Technical, Essex, 1987, pp. 57, 91, 96, 98, 108, 127, 266.
2. Reimers, F., *The Basic Principles for Pharmacopoeial Tests*, Heinemann Medical Books, London, 1956, p. 19.
3. Vogel, A.I. *A Text-Book of Macro and Semimicro Qualitative Inorganic Analysis*, 4th ed., Longmans, Green, London, 1954, p. 42.

3.5   Amines, primary aromatic (European Pharmacopoeia 2.3.1)

This test identifies the substance as a primary aromatic amine of the general structure as shown in Figure 3.5.1 below. Any compound containing this structure could be called a primary aromatic amine, but only some will give a positive reaction in the test. About 25 monographs make reference to amines and they represent a broad spectrum of structurally unrelated compounds.

The test makes use of one of the classic procedures for the distinction between primary, secondary, tertiary, and quaternary amines, reaction with nitrous acid and coupling with β-napthol.

In the first step of analysis, the prescribed solution is acidified with dilute *hydrochloric acid R*, and 0.2 ml of *sodium nitrite solution R* is added. The primary aromatic amine is diazotized under these conditions (Figure 3.5.2).

Having allowed 1 to 2 min to complete the above reaction, 1 ml of *β-napthol solution R* is added in the second step of analysis. This gives an intense orange or red color, and usually a precipitate of the same color, of the condensation product given below (Figure 3.5.3).

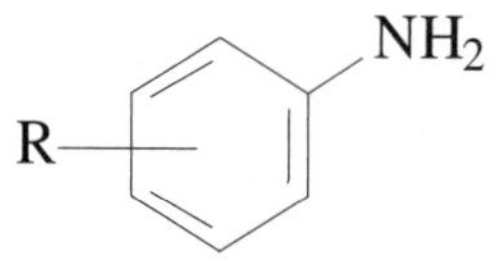

Figure 3.5.1 General primary aromatic amine.

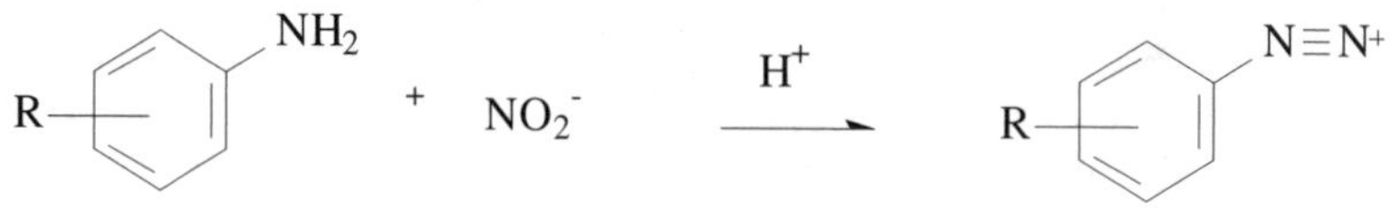

Figure 3.5.2 Diazotization.

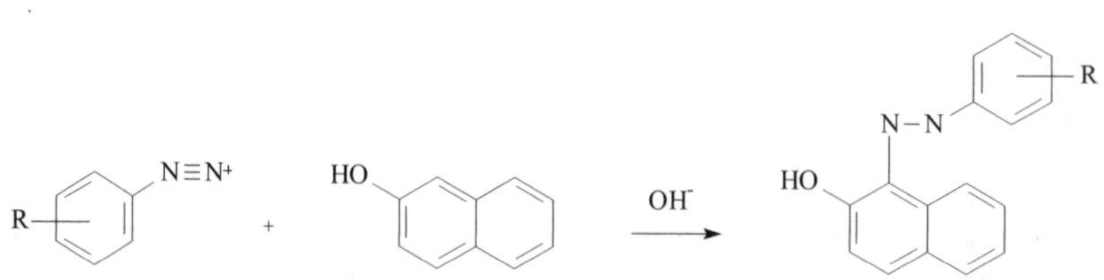

Figure 3.5.3 Condensation with β-napthol.

The selectivity of the test comes mainly from the other amines' lack of reaction toward nitrite. A primary aliphatic amine will not be diazotized under the test conditions given, but instead will be converted into its corresponding alcohol. Nitrogen is liberated in the reaction.

$$RNH_2 + NO_2^- \rightarrow ROH + N_2$$

Both aromatic and aliphatic secondary amines will react by forming N-nitrosamines, as shown below. These are in most cases yellow.

$$RR_1NH + NO_2^- \rightarrow RR_1NNO + H_2O$$

Purely aromatic or purely aliphatic tertiary amines and quaternary ammonium salts do not react with nitrous acid, but mixed aromatic-aliphatic amines of the dimethylaniline type form green or yellow p-nitroso derivatives (Figure 3.5.4) when treated with nitrous acid. This reaction has also been used for the determination of nitrites.

Primary aromatic diamines cannot be diazotized, but other reactions might occur. Thus o-phenylenediamine yields a triazole derivative, and m-phenylenediamine yields an azodye (bismarck brown) by self-coupling if submitted to the test conditions, and this reaction too has been used for the determination of nitrites.[1]

A false negative reaction could be the result if the β-napthol solution R is not capable of creating an alkaline environment in the test solution, since α-nitroso-β-napthol is formed under acidic conditions instead of the intended azo dye (Figure 3.5.5).

Finally, it should be mentioned that not all compounds containing the general structure of that in Figure 3.5.1 should necessarily be expected to give a positive reaction in the test. Other substituents on the aromatic ring of the primary amine, or even the properties of other parts of the molecule in question, might prevent reaction.[2,3]

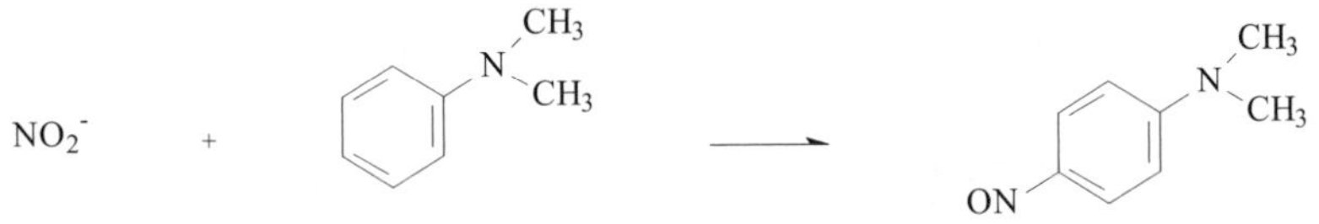

Figure 3.5.4 p-nitroso derivative.

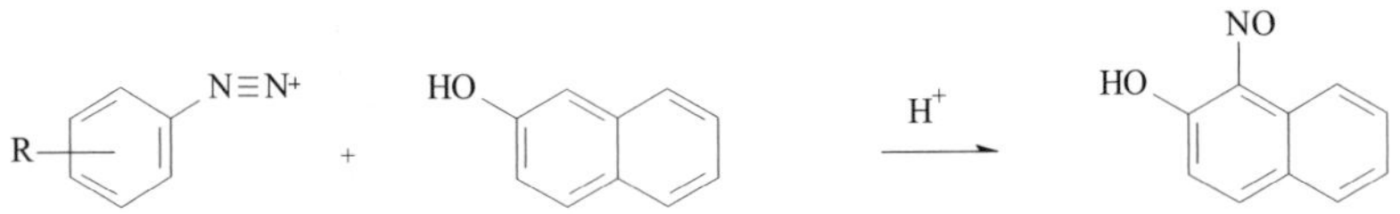

Figure 3.5.5 α-nitroso-β-napthol.

References

1. Lange, B. and Vejdělek, Z.J., *Photometrische analyse*, 1st ed., Verlag Chemie, Wienheim, 1980, p. 369.
2. Vogel, A.I., *Elementary Practical Organic Chemistry*, Part 2, *Qualitative Organic Analysis*, 5th ed., Longmans, 1962, p. 447.
3. Veibel, S., *The Identification of Organic Compounds*, 7th ed. (4th English ed.), G.E.C. GAD Publishers, Copenhagen, 1971, p. 265.

3.6   Ammonium salts (European Pharmacopoeia 2.3.1)

The test identifies the substance to be examined as a salt of ammonium, NH_4^+. At the present only three monographs give reference to ammonium salts, all of which describe simple inorganic salts. Both organic and inorganic ammonium salts are relatively unstable and give off ammonia, especially when heat is applied. Ammonia therefore does not play a role as an indifferent counter-ion. Ammonium is almost identical in size to potassium, and therefore it has properties that are almost identical to this ion. It is a very weak acid with a pKa of 9.25, meaning that solutions have to be fairly alkaline to bring ammonium onto the ammonia form. Like the alkali metals, it forms water-soluble salts with nearly all inorganic anions apart from a few exceptions. One of the exceptions is the precipitate it forms with cobalt nitrite, and this precipitate, together with the volatility of ammonia, forms the basis of the test.

A glass apparatus as shown below (Figure 3.6.1) can be used. The prescribed solution is placed in test tube A, and 0.2 g of *magnesium oxide R* is

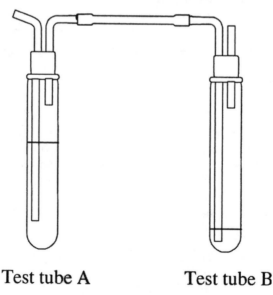

Test tube A   **Test tube B**

Figure 3.6.1 Test tube assembly.

added. A current of air is passed through the mixture, and the gas that escapes is directed to just beneath the surface of the reagent in test tube B. This is a mixture of 1 ml of *0.1 M hydrochloric acid* and 0.05 ml of *methyl red solution R*. The color of the indicator changes to yellow.

The ammonium in the test solution is deprotonized by the magnesium hydroxide.

$$NH_4^+ + OH^- \rightarrow NH_3\uparrow + H_2O$$

Compared to the test conditions described in 3.7. Ammonium salts and salts of volatile bases where sodium hydroxide is used, magnesium affords a much less harsh reaction condition. Being a weaker base than sodium hydroxide, magnesium hydroxide only liberates ammonia from ammonium salts and is not capable of hydrolyzing amides and other nitrogen-bearing organic molecules as is the case in 3.7. Ammonium salts and salts of volatile bases,[1] especially since no heat is applied and because the ammonia gas is guided from the test solution by a current of air through the system. Very unstable organic amides and inorganic nitrogen compounds such as CaN_2 or $NaNH$ would, however, still be hydrolyzed, liberating ammonia.

In test tube B, ammonia gas is captured in the water and neutralized by the hydrochloric acid in the mixture. When all the hydrochloric acid is used, ammonia will change the indicator methyl red from red to yellow.

In the last step of the analysis, 1 ml of a freshly prepared 100 g/l solution of *sodium cobaltnitrite R* is added to the mixture in test tube B, upon which a yellow precipitate of ammonium hexanitritocobaltate(III) is formed.

$$3NH_4^+ + [Co(NO_2)_6]^{3-} \rightarrow (NH_4)_3[Co(NO_2)_6]\downarrow$$

As discussed in 3.27. Potassium, a similar precipitate is formed if potassium is present.

$$3K^+ [Co(NO_2)_6]^{3-} \rightarrow K_3[Co(NO_2)_6]\downarrow$$

There are some disagreements in the literature about which ions interfere with the determination. Some state that potassium, thallium, and rubidium give precipitates similar to the one seen with ammonia, and the reaction is, as mentioned, also used in the test in 3.27. Potassium. One reference lists cesium, barium, zirconium, lead, and mercury as ions that interfere with the test but without stating the nature of interference.[2] However, since none of these cations are capable of traveling from test tube A to test tube B, they should be viewed strictly as possible causes of interference if they unintentionally are present in test tube B and not as candidates of false positive reactions. If the test solution by mistake or contamination is alkaline, the reagent will be destroyed by precipitation of black cobalt(III) hydroxide. The test is also mentioned to be sensitive toward reducing substances.[3]

References

1. Feigl, F., *Tüpfelanalyse*, Band 1, *Anorganisher Teil, vierte deutsche auflage*, Akademishe Verlagsgesellschaft, Frankfurt am Main, 1960, p. 244.
2. Kolthoff, I.M. and Elving, P.J., *Treatise on Analytical Chemistry*, Part I, Vol. 1, Interscience, New York, 1959, pp. 361, 370.
3. Reimers, F., *The Basic Principles for Pharmacopoeial Tests*, Heinemann Medical Books, London, 1956, p. 20.

3.7   Ammonium salts and salts of volatile bases (European Pharmacopoeia 2.3.1)

The test identifies the substance to be examined as either a salt of ammonium, NH_4^+, or as the salt of a volatile (nitrogen) base. At the present only two monographs give reference to ammonium salts and salts of volatile bases. One monograph describes a distillate of bituminous schists (black slate) containing both lower amines and inorganic ammonia. The other is a tertiary amine, which upon hydrolysis yields ammonia. Ammonium is almost identical in size to potassium and therefore has properties that are almost identical to this. Like the alkali metals it forms water-soluble salts with all but a few inorganic anions. A precipitation therefore is not the obvious choice for a method of identification, and consequently the test is based on ammonia's most obvious difference from the alkali metals, its volatility.

In the test, about 20 mg of the substance to be examined is dissolved in 2 ml of *water R*, unless another solution is prescribed. Two ml of dilute *sodium hydroxide solution R* is added. The solution is heated, and by this it gives off vapor that can be identified by its odor and by its alkaline reaction (2.2.4).

The sodium hydroxide brings ammonia onto the volatile basic form, and this gas is recognized by its characteristic smell and its alkaline reaction to litmus paper. Neither the volatility nor the smell differentiates ammonia from some of the smallest organic amines that behave in much the same manner. A possible cause of false positive reactions is nitrogen containing organic compounds, as for example amides, which under the test conditions described might liberate ammonia or a lower amine, as shown here (Figure 3.7.1).

Also certain inorganic substances such as cyanides and CaN_2 or $NaNH$ will hydrolyze in the alkaline environment, liberating ammonia. If it is necessary to differentiate between ammonia and volatile bases, the test 3.6. Ammonium salts should be used, since this test does not give a positive reaction with volatile bases.

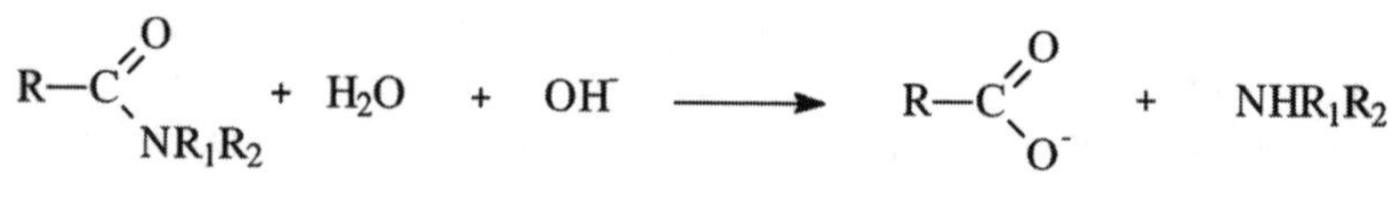

Figure 3.7.1 Liberation of amine.

If the test solution is contaminated with certain cations, for example zinc, they might participate with ammonia in water-soluble complexes. Some of these complexes are strong enough to hinder a positive reaction.[1]

It should be recognized that not only NH_4^+ and protonized organic nitrous bases will give a positive reaction. NH_3 and free bases will also do so, since they are already in volatile form even before sodium hydroxide is added.[2]

References

1. Lamm, C.G. and Rancke-Madsen, E., *Uorganisk Analytisk Kemi*, G.E.C Gads Forlag, København, 1970, p. 36.
2. Reimers, F., *The Basic Principles for Pharmacopoeial Tests*, Heinemann Medical Books, London, 1956, p. 20.

3.8   Antimony (European Pharmacopoeia 2.3.1)

This test identifies the substance to be examined as a salt of antimony(III), Sb^{3+}, or antimony(V), Sb^{5+}. Antimony(III) and antimony(V) were formerly used in the oral treatment of intestinal worms and topically in the treatment of infections of protozoan parasites in the skin. But since especially tetravalent antimony is poisonous, they have generally been replaced by less toxic alternatives. At present, there are no monographs of antimony compounds in the European *Pharmacopoeia*. Antimony forms both tetra- and pentavalent ions, but the pentavalent is mainly found in oxides containing the antimonate ion, SbO_4^{3-}. Antimony(III), on the other hand, can be found as the free dissociated ion, Sb^{3+}, but, as also for example bismuth, since it reacts with water at neutral pH, forming antimonate.

$$Sb^{3+} + H_2O \rightarrow SbO^+ + 2H^+$$

Antimony(III) belongs to the group of amphoteric cations and will react to an increase in pH by forming a white precipitate of hydrated antimony(III), which dissolves upon further addition of base through the formation of antimonites.

$$Sb^{3+} + 6OH^- \rightarrow Sb_2O_3\downarrow + 3H_2O$$

$$Sb_2O_3\downarrow + 2OH^- \rightarrow 2SbO_2^- + H_2O$$

Its ability to form a complex with tartrate and the characteristics of its sulphide salt constitute the bases of the test.

About 10 mg of the substance to be examined is dissolved in a solution of 0.5 g of *sodium potassium tartrate R* in 10 ml of *water R*. When the solution has cooled down, *sodium sulfide solution R* is added dropwise. An orange-red sulfide precipitate is formed regardless of the valence of the antimony present.

$$2Sb^{3+} + 3H_2S \rightarrow Sb_2S_3\downarrow + 6H^+$$

$$2Sb^{5+}\ 5H_2S \rightarrow Sb_2S_5\downarrow + 10H^+$$

In the last step of the analysis, dilute *sodium hydroxide solution R* is added to dissolve the precipitate by forming antimonite and thioantimonite.

$$2Sb_2S_3\downarrow + 4OH^- \rightarrow SbO_2^- + 3SbS_2^- + H_2O$$

The selectivity of the method is given first by the ability to stay dissolved in a solution containing tartrate, second by the color of the sulfide precipitate, and finally by the fact that the sulfide salt dissolves in sodium hydroxide. The first property distinguishes it from bismuth(III) and the other cations forming insoluble oxides in neutral or alkaline solutions. But since the test does not show that a precipitate is formed in pure water, which dissolves when tartrate is added, all water-soluble cations are not excluded. So it should be viewed as a trick to facilitate dissolution only and not a part of the identification. The color of the sulfide precipitate is unique, and it is the most important criterion for a positive identification; if there is any doubt when judging the result, preparing a positive control would be constructive. The solubility of the sulfide salt in sodium hydroxide is a characteristic shared with, for example, the sulfide salt of arsenate, and in classic inorganic separation the sulfide precipitate solubility in hydrochloric acid or polysulfide is used instead.[1-3]

References

1. Vogel, A.I. and Svehla, G., *Vogel's Qualitative Inorganic Analysis*, 6th ed., Longman Scientific & Technical, Essex, 1987, pp. 91, 94.
2. Vogel, A.I., *A Text-Book of Macro and Semimicro Qualitative Inorganic Analysis*, 4th ed., Longmans, Green, London, 1954, p. 246.
3. Charlot, G., *Qualitative Inorganic Analysis, A New Physio-Chemical Approach*, Methuen, London, p. 221.

3.9   Arsenic (European Pharmacopoeia 2.3.1)

The test identifies the substance to be examined as a salt of arsenic(III) or arsenic(V), or a compound containing bound arsenic made available for analysis by chemical degradation. Due to their potential toxicity, arsenic substances must be considered mostly a reminiscence of historical interest. Before World War II a fairly large group of substances containing covalently bound arsenic was used in the treatment of protozoa. One of the most widely known arsenic drugs was arsphenamine (Figure 3.9.1), which was used in

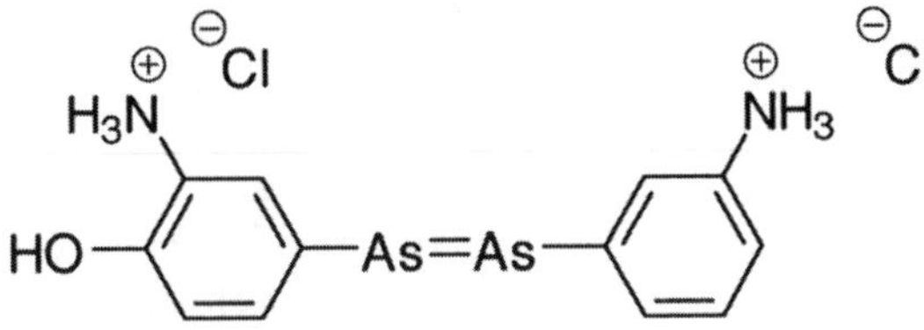

Figure 3.9.1 Arsphenamine.

the treatment of syphilis. It is by some considered as the first modern chemotherapeutic agent. Newer drugs with lesser side effects have now replaced all these compounds, and currently no monograph references arsenic.

Arsenic is in the periodic table placed close to the border between metals and nonmetals and it partly has the properties of both. This means that it can be found either as an ionic species having a valence of +3 or +5, or covalently bound in, for example, an organic structure. As its fellow members in *p*-block group V, nitrogen and phosphor, it is has a great tendency to form oxo and oxo acid compounds. Arsenic(III) is, in analogy to, for example, phosphor, in an alkaline solution found as the oxo anion AsO_3^{3-} and in acidic solution as the acid H_3AsO_3. Arsenic(V) is in the same way found as H_3AsO_4 or AsO_4^{3-} according to pH. Ammonium and alkali salts of arsenic(III) and arsenic(V) are water soluble, whereas the rest are insoluble. All salts are, however, soluble in mineral acid. Solutions of arsenic substances are colorless. Inorganic arsenic compounds are confirmed human carcinogens, trivalent ions being more poisonous than pentavalent. Organic compounds are less toxic.

This very simple test is based on the appearance of metallic arsenic formed by the reduction of arsenic ion by hypophosphorous acid. The same methodology is used in the limit test 6.2. Arsenic, method (b). Five ml of the prescribed solution is heated in a water-bath with an equal volume of *hypophosphorous R*. A brown precipitate is formed.

In the strong hydrochloric acid solution with hypophosphorous acid, both arsenic(III) and arsenic(V) will through arsenic chloride be reduced to amorphous colloidal metallic arsenic as given in this example with arsenic(III):

$$H_3AsO_3 + 3HCl \rightarrow AsCl_3 + 3H_2O$$

$$2AsCl_3 + 3H_3PO_2 + 3H_2O \rightarrow 2As\downarrow + 3H_3PO_3 + 6HCl$$

Any compound that can be reduced by a hydrochloric acid hypophosphorous solution and giving a dark precipitate of a metal will give a positive reaction. The reaction is dealt with in more detail in 6.2. Arsenic, method (b).[1]

Reference

1. Hartke, K., 2.4.2. Arsen (10. Lfg. 1999), in Hartke, K. et al., *Arzneibuch-Kommentar zum Europäische Arzneibuch*, Band I, *Allgemeiner Teil*, Wissenschaftliche Verlagsgellschaft, Stuttgart/Govi-Verlag — Pharmazeutischer Verlag, Escbor.

3.10   Barbiturates, non–nitrogen substituted (European Pharmacopoeia 2.3.1)

This test identifies the substance to be examined as a non–nitrogen substituted barbiturate or thiobarbiturate. Barbiturates are a large group of structurally related compounds with anxiolytic and hypnotic effect. They are derivatives of barbituric acid (Figure 3.10.1) or thiobarbituric acid (Figure 3.10.2).

The various barbiturates and thiobarbiturates (hereafter barbiturates) are different primarily in their alkyl or aryl substituents at the five position. At present less than monographs reference barbiturates, non–nitrogen substituted, all of them describing pharmacologically active barbiturates.

A test first described by Parri in 1924 in which the barbiturate gives a colored complex with cobalt(II) has been adapted by the pharmacopoeia. The test parameters of the determination have been modified several times through the years, but the presence of a number of elements constitutes the essence of the procedure: a M^{2+} ion, normally cobalt(II) or copper(II); an organic solvent, preferably one with a Lewis base character; and an alkaline reaction and the absence of water in the test solution. Furthermore ammonia, or an organic amine, is sometimes added.

If an organic amine or ammonia is present in limited concentrations a tetrahedral complex of the proposed structure (Figure 3.10.3) is formed. The M^{2+} ion, in this example cobalt(II), is the central atom, and two equivalents of barbitutate entities and two equivalents of amine are the ligands.

In the absence of an amine, the solvent molecules of the test solution will serve as the ligand (Figure 3.10.4). In such cases an octahedral complex is more likely the result.[1]

The color intensity of the complex (Figure 3.10.4) is much smaller than the color intensity of complex (Figure 3.10.3), so in most assays based on the Parri methodology, an amine is added. Other structures of the complexes have been suggested, for example one in which one of the carbonyl groups adjacent to the nitrogen in the barbituric acid ring coordinates to the cobalt ion.

Several organic solvents have been suggested for the procedure, but methanol and alcohol are among the most widely used. This is most likely because it has a relatively strong Lewis base and therefore has a strong

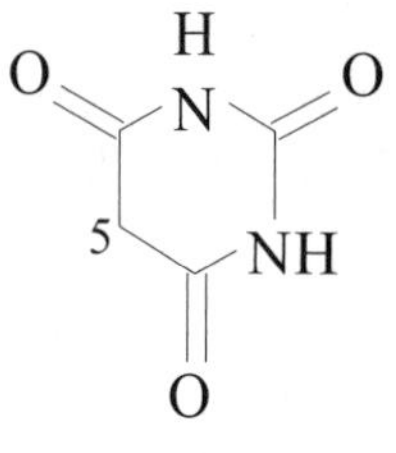

Figure 3.10.1 Barbituriz acid. 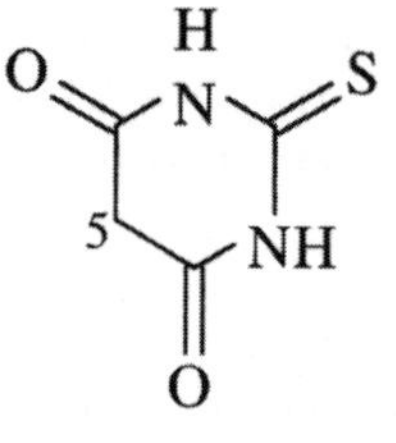   *Figure 3.10.2* Thiobarbituriz acid.

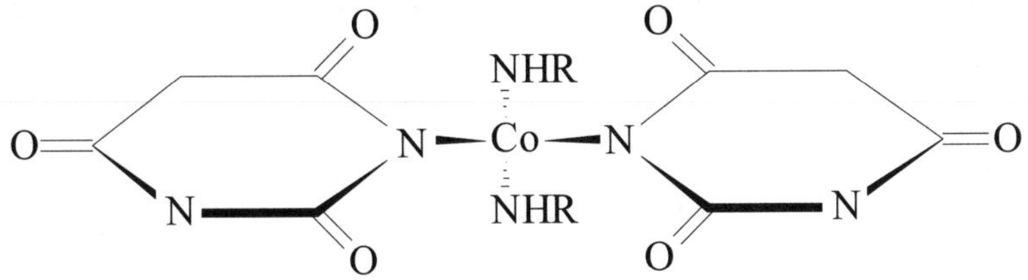

Figure 3.10.3 Amine complex.

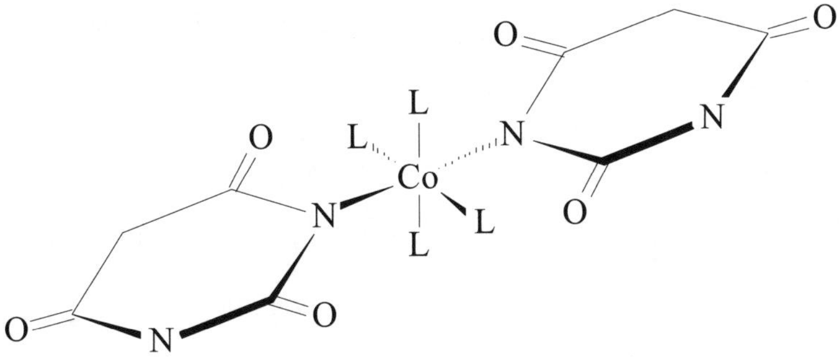

Figure 3.10.4 Solvent ligand complex.

affinity to cobalt. If water is present at a level higher then about 0.1%, the water molecules will replace the other ligands, giving a colorless complex.

The test solution must be alkaline in order to deprotonate the barbituric acid nitrogen. Many different bases, including ammonia and various inorganic hydroxides, have been suggested. The various bases have a marked effect on the stability of the complex formed, and isopropylamine is claimed to be the best choice. Also many different amines have been tested. Besides ammonia, piperazine, pyridine, piperadine, isopropylamine, low chain alkyl amines and cyclohexylamine could be mentioned. Obviously some of the nitrogen bearing bases act both as a base and as amine ligand. All these variations give different selectivity and sensitivity, of importance especially when using the methodology in quantitative work.

A test solution is prepared by dissolving about 5 mg of the substance to be examined in 3 ml of *methanol R*. A 0.1 ml solution containing 100 g/l of *cobalt nitrate R* and 100 g/l of *calcium chloride R* is added. After mixing, 0.1-ml of dilute *sodium hydroxide solution R* is added with shaking. A violet-blue color and precipitate are formed.

In the variant of the procedure adapted by the *European Pharmacopoeia*, no amine ligand is added. Instead, calcium is added, and this gives a variant of the above complexes of a structure, which has not been elucidated.[2] The complex is insoluble in methanol. The presence of calcium also explains the violet-blue color, since most of the complexes with amines are violet. The cobalt complexes are more likely to precipitate when making the test solution

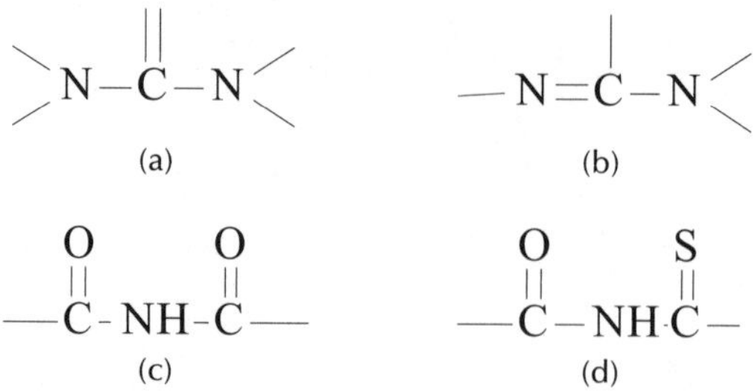

$$>N-\overset{\overset{\displaystyle \|}{O}}{C}-N<\qquad\qquad -N=\overset{\overset{\displaystyle |}{}}{C}-N<$$

(a)   (b)

$$-\overset{\overset{\displaystyle O}{\|}}{C}-NH-\overset{\overset{\displaystyle O}{\|}}{C}-\qquad\qquad -\overset{\overset{\displaystyle O}{\|}}{C}-NH\cdot\overset{\overset{\displaystyle S}{\|}}{C}-$$

(c)   (d)

Figure 3.10.5 Functional groups.

strongly alkaline or by heating. This was used in the first variant of the method in *European Pharmacopoeia* first edition, in which the test solution is heated and the precipitate is studied under the microscope to reveal the exact barbiturate under examination.[3] The intensity of color produced apparently correlates somewhat to the nature of the substituents carbon nr. 5 (Figure 3.10.27).[4]

Barbiturates that are substituted at both ring nitrogen atoms in the ring are not capable of forming a complex and give no color reaction. Barbiturates that are substituted at one of the nitrogens can only form complexes, but these are not very stable.[5]

The procedure is not entirely specific. Some authors state that any compound containing one of the functional groups (Figures 9.10.5a–f) might give a false positive reaction.[6] Besides, a number of substances are capable of giving colored complexes under alkaline conditions: phenols and 1,2-diphenols, polyhydroxy compounds, dipeptides and nitrogen- containing heterocyclic systems, carboxylic acid amides and imides.[7–9]

References

1. Wolfram, H. and Umlauf, S., Barbituratanalytik mit schwermetallionen von den anfängen bis zur gegenwart, *Pharmazie in Unserer Zeit*, 22, 207, 1993.
2. Pesez, M.M., Sur une nouvelle méthode d'identification des dérivés barbituriques: applications, *J. Pharm. Chimie*, 22, 69, 1938.
3. Hartke, K., 2.3.1. Identitätsreaktionen (10. Lfg. 1999), in *Arzneibuch-Kommentar zum Europäische Arzneibuch*, Band I *Allgemeiner Teil*, Hartke, K. et al., Wissenschaftliche Verlagsgellschaft, Stuttgart/Govi-Verlag — Pharmazeutischer Verlag, Escbor.
4. Bult, A., The cobaltous amine reaction 1. The behaviour of the 5,5-disubstituted barbituric acids, *Pharmaceutisch Weekblad*, 111, 157, 1976.
5. Kakáč, B. and Vejdělek, Z. J., *Handbuch der Photometrishe Analyse Organisher Verbindungen*, Band 2, Verlag Chemie, Weinheim, 1974, p. 802.
6. Pesez, M. and Bartos, J., *Colometric and Fluorimetric Analysis of Organic Compounds and Drugs*, Marcel Dekker, New York, 1974, p. 387.
7. Kakáč, B. and Vejdělek, Z.J., *Handbuch der Photometrishe Analyse Organisher Verbindungen*, 2. Ergänzungsband, Verlag Chemie, Weinheim, 1983, p. 350.

8. Vejdělek, Z.J. and Kakáč, B., *Farbreaktionen in der Spectrophotometrischen Analyse Organischer Verbindungen*, Band II, VEB Gustav Fischer Verlag, Jena, 1973, p. 396.
9. Lange, B. and Vejdělek, Z.J., *Photometrische Analyse*, 1st ed., Verlag Chemie, Wienheim, 1980, p. 430.

3.11 Benzoates (European Pharmacopoeia 2.3.1)

The test identifies the substance to be examined as either a benzoate salt or a substance that yields benzoic acid upon chemical manipulation. Reference to benzoates is at present made in only three monographs describing benzoic acid (Figure 3.11.1) itself, its sodium salt, and benzoyl peroxide, which upon basic hydrolysis yields two benzoic acid equivalents. Being a carboxylic acid, benzoic acid has a water solubility that depends strongly on the pH of the solution. The free acid is sparingly soluble in cold water but significantly more soluble in hot water. It is soluble in alcohol and ether. All common benzoate salts except the salts of silver and basic iron(III) are water soluble.

There are three identification tests for benzoates in the European *Pharmacopoeia*, test (a) in which benzoate is precipitated with a ferric salt to get the above-mentioned insoluble iron(III) salt, and test (b) and test (c), where the physical properties of benzoic acid are used for identification.

Test (a)

This test utilizes the ability of iron(III) to form colored complexes and precipitates.

To 1 ml of the prescribed solution add 0.5 ml of *ferric chloride solution R1*. A dull-yellow precipitate of basic iron(III) benzoate, soluble in *ether R*, is formed.

$$3C_6H_3COO^- + 2Fe^{3+} + 3H_2O \rightarrow (C_6H_5COO)_3Fe.Fe(OH)_3\downarrow + 3H^+$$

The precipitation takes place from neutral solutions, but the precipitate dissolves in acid environment as the benzoate salt is converted into the free acid. The selectivity and robustness of the precipitation suffers in that a rather high number of organic and inorganic compounds give colored complexes

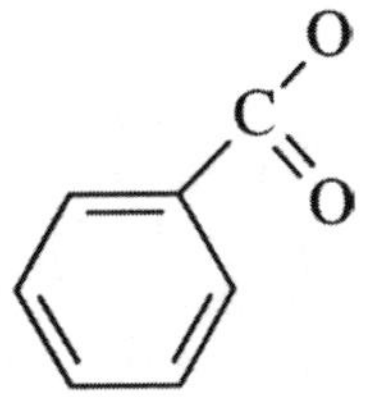

Figure 3.11.1 Benzoic acid.

or precipitates with iron(III). Examples of determinations of various organic substances can be found in the literature.[1,2]

Test (b)

The acid form of benzoic acid has a relatively low melting point and high vapor pressure, and this is used in test (b).

Place 0.2 g of the substance to be examined (treated if necessary as prescribed) in a test tube. It is moistened with 0.2 to 0.3 ml of *sulfuric acid R*, and the bottom of the test tube is gently warmed. A white sublimate is deposited on the inner wall of the tube.

Adding the sulfuric acid to the test substance liberates the benzoic acid from its salt (if present as a salt) and brings it to its volatile acid form. When heat is applied some of the benzoic acid will evaporate without melting first and condense on the inner wall of the cooler part of the test tube. The process of evaporating from the solid phase is called sublimation.

The specificity of the test comes from two properties of benzoic acid: its chemical stability and its vapor pressure. Benzoic acid does not char when treated with concentrated sulfuric acid, as many organic substances do. One reference mentions, however, that heat should be applied carefully since charring will otherwise initiate.[3]

Test (c)

Benzoic acid's water solubility dependency of test solution pH and the melting point of the acid form of test (c).

Dissolve 0.5 g of the substance to be examined in 10 ml of *water R*, or 10 ml of the prescribed solution is used. Add 0.5 ml of *hydrochloric acid R*. The precipitate obtained, after crystallization from warm *water R* and drying *in vacuo*, has a melting point of 120 to 124°C.

The relevance of a melting point determination in the case of benzoic acid is that it has a quite narrow melting range, and this is actually in the *European Pharmacopoeia*'s general monograph for thermometers. The melting point obtained, however, is a function of the purity of the substance examined. For this reason, a recrystallization is indicated. Impure substances will most often give a lower than expected melting point and a wider melting range.

References

1. Vejdělek, Z.J. and Kakáč, B., *Farbreaktionen in der Spectrophotometrischen Analyse Organischer Verbindungen*, Band II, VEB Gustav Fischer Verlag, Jena, 1973, p. 351.
2. Vogel, A.I., *A Text-Book of Macro and Semimicro Qualitative Inorganic Analysis*, 4th ed., Longmans, Green, London, 1954, p. 410.
3. Hartke, K., 2.3.1. Identitätsreaktionen (10. Lfg. 1999), in *Arzneibuch-Kommentar zum Europäische Arzneibuch*, Band I, *Allgemeiner Teil*, Hartke, K. et al., Wissenschaftliche Verlagsgellschaft, Stuttgart/Govi-Verlag — Pharmazeutischer Verlag, Escbor.

3.12   Bismuth (European Pharmacopoeia 2.3.1)

The test identifies the substance to be examined as a salt of bismuth, Bi^{3+}. Several insoluble bismuth salts have found use orally as antacids and topically in the treatment of skin disorders. Others have been used in the treatment of syphilis and as an amoebicide. Most of these therapeutic substances have been replaced by more modern alternatives, and reference to bismuth is currently made in only four monographs, and just one dictate method (a). Bismuth forms trivalent and pentavalent ions, but the trivalent is by far the most common. Most bismuth(III) and bismuth(V) forms insoluble salts, especially under alkaline conditions. This behavior and the insolubility of bismuth sulphide form the basis of test (a). In test (b) a colored complex between bismuth and thiourea is formed.

Test (a)

To 0.5 mg of the substance to be examined add 10 ml of dilute *hydrochloric acid R* or use 10 ml of the prescribed solution. Heat to boiling for 1 min. Cool and filter if necessary. This procedure should bring the most likely otherwise insoluble bismuth salt into solution. In the next step of analysis, 1 ml of the solution obtained is added to 20 ml of water R, upon which a white or slightly yellow precipitate is formed. Bismuth(III) hydrolyzes water, forming the bismuthyl ion.

$$Bi^{3+} + H_2O \rightarrow BiO^+ + 2H^+$$

This ion forms an insoluble salt with most anions. In the case of a diluted hydrochloric acid solution, the precipitate will predominantly be with chloride, but in the case of other anions a mix of salts could be present: $BiOCl$, $(BiO)_2SO_4$, and $Bi(OH)_2NO_3$.

Finally, 0.05 ml to 0.1 ml of *sodium sulphide solution R* is added, upon which the precipitate should turn brown, owing to the formation of bismuth sulfide.

$$2BiOCl + 3S^- \rightarrow Bi_2S_3\downarrow$$

Since bismuth sulfide is the only brown sulfide, a high degree of selectivity is obtained, although several cations are soluble in hydrochloric acid and precipitate upon dilution with water. The exact color of the various sulfide salts is the subject of some debate, though. Some references claim that bismuth sulfide is black, and some state that tin(II) sulfide is brown.[1,2]

Test (b)

To about 45 mg of the substance to be examined add 10 ml of *dilute nitric acid R* or use 10 ml of the prescribed solution. Boil for 1 min. Allow to cool and filter if necessary. As in test (a), this should bring the bismuth salt in question into solution. In the second step, to 5 ml of the solution obtained is added 2 ml of a 100 g/l solution of *thiourea R*. By this, a yellowish-orange color or orange precipitate or complex is formed.

Cadmium, copper, mercury, silver, and tin form white precipitates with thiourea when present in high concentrations. Only antimony(III) (and vanadate) gives a weak yellow color with thiourea. Hg^+, SeO_3^{2-} and SeO_4^{2-} are reduced and interfere by forming colored complexes and precipitates.[3]

In the last step of analysis, add 4 ml of a 25 g/l solution of *sodium fluoride R*, and this must not result in a decolorization of the test. If the color developed in step two of the analysis was caused by the presence of antimony(III), this will be destroyed by the presence of the fluoride ion that affects the antimony(III) complex but not the bismuth complex.[4]

References

1. Hildebrand, W.F. et al., *Applied Inorganic Analysis*, John Wiley, New York, 1953, p. 232.
2. Kodama, K., *Methods of Quantitative Inorganic Analysis*, Interscience Publishers, New York, 1963, p. 165.
3. International Union of Pure and Applied Chemistry, *Reagents and Reactions for Qualitative Inorganic Analysis*, fifth report, Butterworths, London, 1964.
4. Kolthoff, I.M. and Elving, P.J., *Treatise on Analytical Chemistry*, Part 2, Vol. 8, 1963, p. 147.

3.13   Bromides (European Pharmacopoeia 2.3.1)

The test identifies the substance to be examined as a salt of bromide (Br⁻) or a substance containing chemically bound bromine that can be released as bromide upon treatment. Reference to bromides is made in about 20 monographs. Test (a) is dictated in the majority of the monographs, and test (b) is in only one case dictated alone. Most of the monographs describe bromide salts, ranging from simple inorganic salt to hydrobromides of complex nitrogen bases, and the rest describe organic substances having bromine chemically attached. In these cases, bromide is released by heating the substance to be examined with anhydrous sodium carbonate to prepare the material for testing. Retrospectively, hydrobromic acid was the third most frequently used acid for forming salts of organic bases, but it has largely been substituted by other anions due to the toxicity of the bromide ion. Most simple inorganic salts of bromide are water soluble. Exceptions are, as for chloride: lead, thallium, silver, mercury, and copper. In general, however, bromide salts are less soluble than the corresponding chloride salts.

Test (a) is a silver salt precipitation of limited selectivity; test (b) is based on bromide's reaction toward the organic reagent fuchsin.

Test (a)

In the test bromide is precipitated as silver bromide, and the precipitate's reaction toward nitric acid and ammonia is used to rule out other silver precipitates. Since a very high number of elements and also organic substances give a precipitate with silver nitrate, the description of selectivity below is not meant to be conclusive. It mainly deals with the other halogens and a few other inorganic cations. Precipitation with silver nitrate is also used for test (a) of 3.16. Chloride and test (a) of 3.19. Iodide, so the difference between the tests of these is dealt with in more detail.

Dissolve in 2 ml of *water R* a quantity of the substance to be examined equivalent to about 3 mg of bromide (Br⁻), or use 2 ml of the prescribed solution. Acidify with dilute *nitric acid R* and add 0.4 ml of *silver nitrate solution R1*. Shake and allow to stand. A curdled, pale yellow precipitate of silver bromide is formed.

$$Ag^+ + Br^- \rightarrow AgBr\downarrow$$

In the next step of analysis the precipitate is centrifuged and washed with three quantities, each of 1 ml, of *water R*. Carry out this operation rapidly in subdued light disregarding the fact that the supernatant solution may not become perfectly clear. This precaution is necessary since the precipitate when exposed to light turns grayish or black as Ag^+ is reduced to the metallic silver.[1]

$$2AgBr\downarrow + \lambda \rightarrow 2Ag\downarrow + Br_2$$

Suspend the precipitate obtained in 2 ml of *water R* and add 1.5 ml of *ammonia R*. The precipitate dissolves with difficulty. This is due to the silver ions' ability to form the diammninoargentate complex.

$$AgBr\downarrow + 2NH_3 \rightarrow [Ag(NH_3)_2]^- + Br^-$$

The precipitate has to be separated from the original test solution since the ions present here, e.g., the cations of the substance to be examined, can disturb the complexation.[2] The solubility of silver bromides is intermediate compared to that of silver chloride and silver iodide. So the complex formation constant of the diammninoargentate complex overrules the solubility constant of silver chloride easily, and with difficulty the solubility constant of silver bromide, but it is not able to do so in the case of silver iodide.

As with chloride and iodide, there is not total agreement in references on which ions are capable of giving false positive reactions, but perhaps the greatest disadvantage of the method is not its relative high numbers of candidates for false positive reactions but rather its sensitivity toward interference. The high number of insoluble silver salts lowers the ruggedness of the method, since the presence of even low concentrations of organic and inorganic cations gives precipitates of various colors.[3,4]

Test (b)

This test is based on bromide's redox characteristic and the ability of bromine to give a color reaction with a solution of decolorized fuchsin. This reagent is also known as malachite green and by many authors as Schiff's reagent, owing to its use in a determination of aldehydes discovered by H. Schiff.

The decolorized fuchsin solution is a 0.1% acidic solution of basic fuchsin, which is a mixture of rosaniline hydrochloride (Figure 3.13.1) and para-rosaniline hydrochloride (Figure 3.13.2).

Other mixtures are sold containing derivatives with more methyl substituents, but they are less suitable for analytical work. This solution is decolorized by adding sodium sulfite, thereby changing Figures 3.13.1 and 3.13.2 into the colorless Figure 3.13.3.[5]

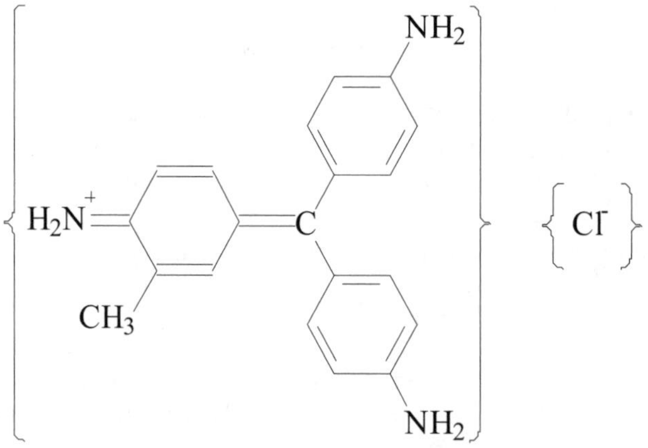

Figure 3.13.1 Rosaline hydrochloride.

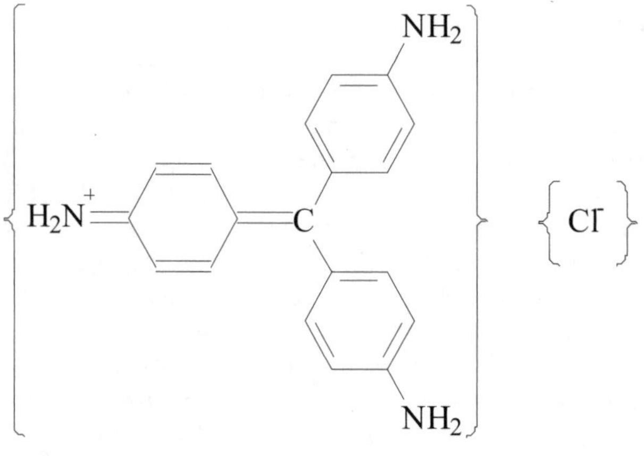

Figure 3.13.2 Para rosaline hydrochloride.

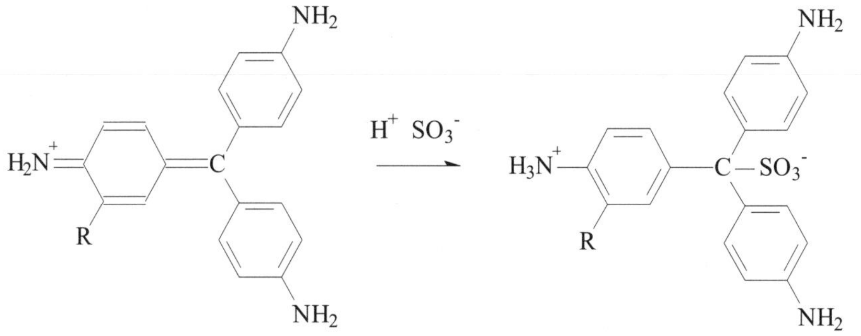

Figure 3.13.3 Decolorization.

Older references often give a structure where the sulfite group is not attached to the central carbon but to one of the ammonium groups. The decolorization is disturbed by the presence of higher than intended concentrations of alkali metals.[6] The reagent must be stored protected from light.

In the first step of analysis, a quantity of the substance to be examined equivalent to about 5 mg of bromide (Br$^-$), or the prescribed quantity, is introduced into a small test tube. To this 0.25 ml of *water R*, about 75 mg of *lead dioxide R*, and 0.25 ml of *acetic acid R* are added, and the tube is shaken gently. In this procedure bromide is oxidized by the lead dioxide to bromine.

$$2Br^- + PbO_2 + 4H^+ \rightarrow Br_2 + Pb^{2+} + 2H_2O$$

In the second step, the inside of the upper part of the test tube is dried with a piece of filter paper and allowed to stand for 5 min. A strip of suitable filter paper of appropriate size is impregnated by capillary, by dipping the tip into a drop of decolorized *fuchsin solution R*. The impregnated part is introduced immediately into the tube. Starting from the tip, a violet color appears within 10 s that is clearly distinguishable from the red color of fuchsin, which may be visible on a small area at the top of the impregnated part of the paper strip. The violet color, which in some references[6] is stated to be blue, is due to pentabromine rosaline (Figure 3.13.4) and hexabromine rosaline (Figure 3.13.5) produced by the addition of bromine to fuchsin.

As stated in the test some fuchsin is produced in the reaction as well. If the impregnated filter paper is exposed to heat, the red color (pink) of fuchsin is restored. When the reagent is used for identifying and determining aldehydes, a deep red-violet addition product is formed. Selectivity is given by the fact that neither chlorine nor iodide colorizes decolorized fuchsin, and secondarily because only bromine but not chlorine is oxidized by lead oxide. The selectivity toward iodide and chloride is so good that the method can be used for the detection of even small amounts of bromides in chlorides and iodides. A variant of the test where the relatively gentle oxidant *lead dioxide* is replaced by concentrated chromic acid can be used for detecting

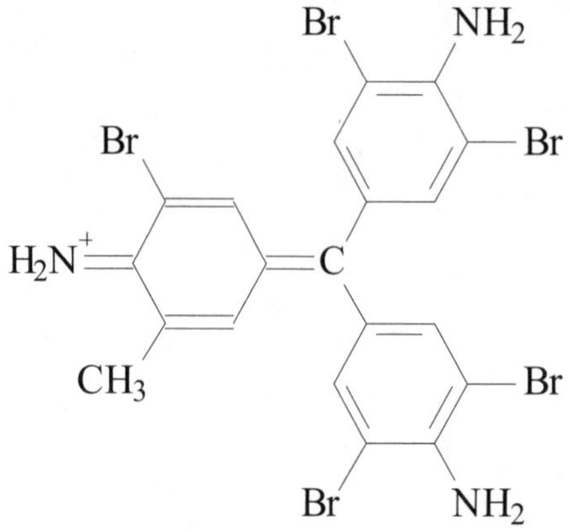

Figure 3.13.4 Pentabromine rosaline.

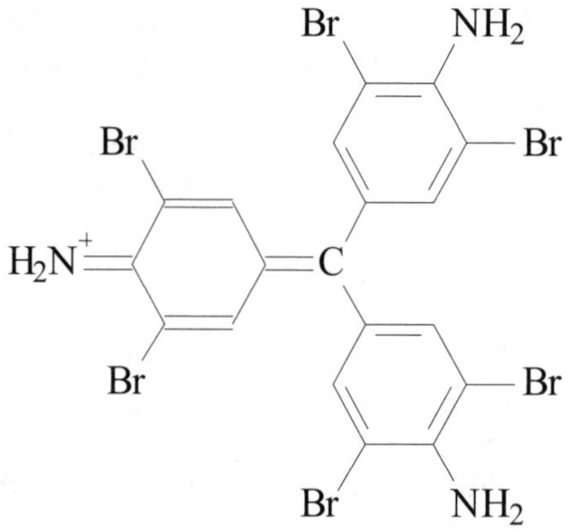

Figure 3.13.5 Hexabromine rosaline.

bromine in organic compounds, even if it is substituted with other halogens as well.

References

1. Vogel, A.I. and Svehla, G., *Vogel's Qualitative Inorganic Analysis*, 6th ed., Long-man Scientific & Technical, Essex, 1987, pp. 57, 63, 66, 76, 176, 246–249.
2. Hartke, K., 2.3.1. Identitätsreaktionen (10.Lfg.1999), in *Arzneibuch-Kommentar zum Europäische Arzneibuch*, Band I, *Allgemeiner Teil*, Hartke, K. et al., Wissen-schaftliche Verlagsgellschaft, Stuttgart/Govi-Verlag — Pharmazeutischer Verlag, Escbor.
3. Kolthoff, I.M. and Elving, P.J., *Treatise on Analytical Chemistry*, Part II, Vol. 7, Interscience Publishers, New York, 1961, p. 361.

4. Charlot, G., *Qualitative Inorganic Analysis. A New Physio-Chemical Approach*, Methuen, London, 1954, p. 259.
5. Rose, U., Duvaux, S., and Fuchs, J., Fuchsin: examination of the composition and use of a classical pharmacopoeial reagent, *Pharmeuropa*, June 1997, Council of Europe, Strasbourg, 1997, p. 338.
6. Feigl, F., *Qualitative Analysis by Spot Tests*, 3th ed., Elsevier New York, 1947, p. 234.

3.14   Calcium (European Pharmacopoeia 2.3.1)

The test identifies the substance examined as a salt of calcium, Ca^{2+}. Reference to calcium is made in about 30 monographs. Owing to the tendency of calcium to form insoluble salts, it is not a common salt of choice for active pharmaceutical ingredients, so the majority of the monographs are for inactive excipients. Among the exceptions are the calcium salts of several amino acids. Being an alkaline earth metal, calcium is found only in the valance two form. Its hydroxide salt is relatively water soluble, whereas the carbonate, sulfate, and phosphate salts are insoluble. Its reaction toward the organic reagent glyoxal-hydroxyanil and the inorganic reagent ferrocyanide is used in the two tests of the pharmacopoeia.

Test (a)

A neutral test solution is prepared containing a quantity of the substance to be examined equivalent to about 0.2 mg of calcium (Ca^{2+}), or the prescribed solution is used. To 0.2 ml of this solution is added 0.5 ml of a 2 g/l solution of *glyoxal-hydroxyanil R* in *alcohol R*, 0.2 ml of dilute *sodium hydroxide solution R*, and 0.2 ml of *sodium carbonate solution R*.

A number of cations will under these circumstances give a red complex with the glyoxal-hydroxyanil reagent as shown in Figure 3.14.1 below.

Of the other alkali earth metals, barium and strontium will give such a complex. But they will reveal themselves by precipitating in the above procedure, as, by the addition of sodium carbonate, they will give a precipitate of insoluble carbonate.

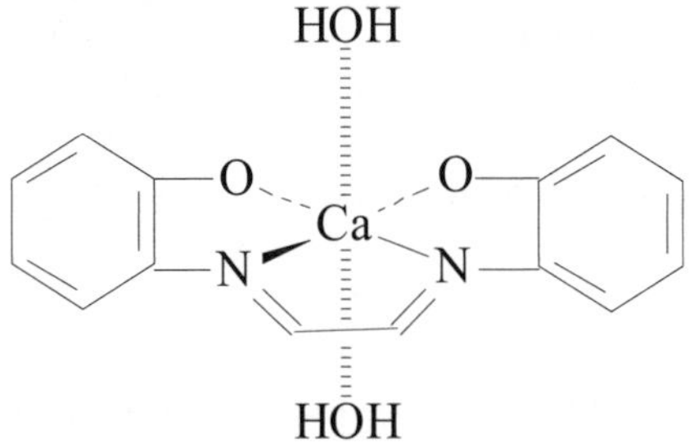

Figure 3.14.1 Glyoxal-hydroxyanil.

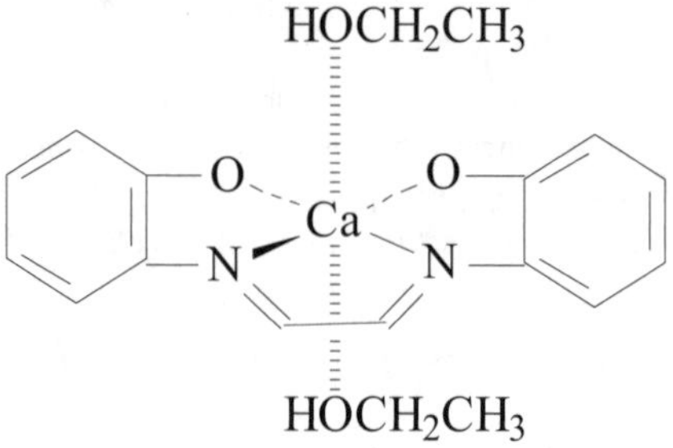

Figure 3.14.2 Complex.

In the next step, the test solution is shaken with 1 ml to 2 ml *chloroform R* and 1 ml to 2 ml of *water R*. The lower chloroform phase is colored red by the formation of the complex (Figure 3.14.2) pictured here.

Besides the earth alkali metals mentioned, a positive reaction in the first step of analysis will be given by cadmium, cobalt, copper, manganese, nickel, zinc, and uranium(VI). But of these only the uranium(VI) can be extracted into chloroform in the second step of the analysis. The extraction is possible since the two water molecules in the calcium complex can by replaced by two ethanol molecules. This complex is soluble in chloroform.

A number of anions are mentioned to cause interference. The complex formation can be eliminated completely by the presence of EDTA, oxalate, fluoride, and phosphate, the organic anions through complex formation with calcium, the inorganic anions through precipitation of calcium. Citrate and carbonate, tartrate, and borate also cause disturbances but only when present in high concentrations.

Since the above methodology has been used in various procedures for quantitative determinations, they have been thoroughly investigated. A good place to start a literature search is the references given below.[1,2]

Test (b)

A test solution is made by dissolving about 20 mg, or the prescribed quantity, of the substance to be examined in 5 ml of *acetic acid R*. The solution should remain clear, when in the first step of the analysis, 0.5 ml of *potassium ferrocyanide solution R* is added. In the second step, a white crystalline precipitate is formed upon the addition of 50 mg of *ammonium chloride R*.

$$Ca^{2+} + 2NH^+ + [Fe(CN)_6]^{4-} \rightarrow (NH_4)_2Ca[Fe(CN)_6]\downarrow$$

At higher calcium and reagent concentrations, the analogue potassium salt precipitates.

$$Ca^{2+} + 2K^+ + [Fe(CN)_6]^{4-} \rightarrow K_2Ca[Fe(CN)_6]\downarrow$$

Magnesium is claimed to behave analogously, also when alcohol is added. Apparently barium can also give a positive reaction if given in high concentrations.[3-5]

References

1. Lange, B. and Vejdělek, Z.J., *Photometrische Analyse*, Verlag Chemie, Wienheim, 1980, p. 80.
2. Fries, J. and Getrost, H., *Organische Reagenzien für die Spurenanalyse*, E. Merck, Darmstadt, 1975, p. 96.
3. Hartke, K., 2.3.1. Identitätsreaktionen (10. Lfg. 1999), in *Arzneibuch-Kommentar zum Europäische Arzneibuch*, Band I, *Allgemeiner Teil*, Hartke, K. et al., Wissenschaftliche Verlagsgellschaft, Stuttgart/Govi-Verlag — Pharmazeutischer Verlag, Escbor.
4. Feigl, F., *Qualitative Analysis by Spot Tests*, 3rd ed., Elsevier, New York, 1947, p. 169.
5. Vogel, A.I. and Svehla, G., *Vogel's Qualitative Inorganic Analysis*, 6th ed., Longman Scientific & Technical, Essex, 1987, p. 137.

3.15 Carbonates and bicarbonates (European Pharmacopoeia 2.3.1)

The test identifies the presence of carbonate, CO_3^{2-}, or bicarbonate, HCO_3^-, in the substance to be examined. Reference to carbonates and bicarbonates is made in less than five monographs, all describing simple inorganic carbonate and bicarbonate salts. The carbonate ion forms insoluble salts with nearly all cations except the alkali metals and ammonia, and the bicarbonate ion, more commonly referred to as hydrogen carbonate, generally has an even lower solubility. Hydrogen carbonate is a weak acid and is unstable in an acidic environment. This characteristic and the insolubility of barium carbonate are used in the identification test.

Suspend or dissolve 0.1 g of the substance to be examined in 2 ml of *water R* in test tube A, as shown in Figure 3.15.1 below. Add 3 ml *dilute acetic acid R*, and the tubing is connected immediately.

The solution or suspension becomes effervescent owing to the instability of the carbonate ion.

$$CO_3^{2-} + 2H^+ \rightarrow CO_2\uparrow + H_2O$$

Upon gentle heating, the gas is led to test tube B containing 5 ml of *barium hydroxide solution R*, and a white precipitate of barium carbonate is formed.

$$CO_2 + Ba^{2+} + 2OH^- \rightarrow BaCO_3\downarrow + H_2O$$

Finally, this precipitate is dissolved upon the addition of an excess of *hydrochloric acid R1*, thereby repeating the first reaction. The precipitate will,

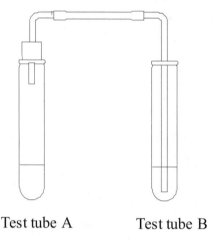

Test tube A   Test tube B

Figure 3.15.1 Test tube assembly.

however, also redissolve if by mistake too much carbon dioxide is passed into test tube B.

$$BaCO_3\downarrow + CO_2 + H_2O \rightarrow Ba^{2+} + 2HCO_3^-$$

The excess carbon dioxide will lower the test solution pH, and this will release the carbonate by changing it to hydrogen carbonate, which does not form an insoluble salt with barium.

The effervescence of gas cannot be regarded as a discriminatory part of the identification, since compounds other than carbonates might liberate CO_2 upon heating. This can be seen even in less obvious examples such as para ammino salicylic acid. Only a few carbonates, such as for example $MgCO_3$ and $FeCO_3$, require heat to be applied for the cleavage to occur; the others will evolve gas as a result merely of contact with the acid. Cyanates will react in the test in the same way as carbonates, releasing CO_2 by the reaction.

$$OCN- + 2H^+ + H_2O \rightarrow CO_2\uparrow + NH_4$$

Sulfite will also give a false positive reaction, since upon treatment with hydrochloric acid it releases sulfuric dioxide, which in test tube B produces a white precipitate of barium sulphite.

$$SO_3^{2-} + 2H^+ \rightarrow SO_2\uparrow + H_2O$$

$$SO_2 + Ba^{2+} 2OH^- \rightarrow BaSO_3\downarrow + H_2O$$

In such a case, though, most likely the smell of sulphur dioxide will reveal the mistake. The other cations giving white precipitates with barium could disturb the analysis if present in test tube B through contamination.[1-3]

References

1. Vogel, A.I. and Svehla, G., *Vogel's Qualitative Inorganic Analysis*, 6th ed., Longman Scientific & Technical, Essex, 1987, p. 150.
2. Reimers, F., *The Basic Principles for Pharmacopoeial Tests*, Heinemann Medical Books, London, 1956, p. 21.
3. Charlot, G., *Qualitative Inorganic Analysis. A New Physio-Chemical Approach*, Methuen, London, 1954, p. 309.

3.16   Chlorides (European Pharmacopoeia 2.3.1)

The test identifies the substance to be examined as a salt of chloride (Cl-) or as a substance containing bound chlorine, which by manipulation releases chloride. Reference to chloride is made in several hundred monographs. The substances they describe span from simple inorganic salts, like sodium chloride, to chloride salts of complex organic cations. A large group is the chloride salts of nitrogen bases, the hydrochlorides. In these, and in most other cases, chloride plays the role of being an indifferent counter-ion. Like the other halogens, chlorine is exclusively found as a monovalent anion. Elemental chlorine is fairly unstable and easily oxidized to the ionic state. Chlorine forms water-soluble salts with almost all inorganic cations, with the exception of lead, thallium, silver, mercury, and copper. In addition to this, chloride can be brought to precipitate with a few less common cations (see 3.30 Silver). In most monographs, sample preparation is a simple dilution in water, but in some cases where chlorine is tightly bound to an organic moiety, for example as in the case of diazepam, chloride has to be released by heating the substance with anhydrous sodium carbonate.

Test (a) is a silver salt precipitation of limited selectivity, and test (b) is a far more specific test that utilizes the chemistry normally used for the identification of chromium.

Test (a)

In the test, chloride is precipitated as silver chloride, and the precipitates reaction toward nitric acid and ammonia is used to rule out other silver precipitates. Since a very high number of elements and also organic substances give a precipitate with silver nitrate, the description below of selectivity is not meant to be conclusive. It mainly deals with the other halogens and a few inorganic cations. Precipitation with silver nitrate is also used for the identification of iodide and bromide, and the difference between the tests of these is therefore dealt with in more detail.

A test solution is prepared by dissolving in 2 ml of *water R* a quantity of the substance to be examined equivalent to about 2 mg of chloride (Cl-), or 2 ml of the prescribed solution is used. Acidify with dilute *nitric acid R* and add 0.4 ml of *silver nitrate solution R1*. Shake and allow to stand. A curdled white precipitate of silver chloride is formed.

$$Ag^+ + Cl^- \rightarrow AgCl\downarrow$$

Since the precipitation is done from a test solution containing nitric acid, a large number of substances otherwise giving reactions can be excluded. One group is the silver salts of weak acids that are unstable in dilute nitric acid and decompose to gas, such as carbonate and sulfite. Still, different substances such as cyanide and several alkaloids can precipitate under these conditions. Chloride, contrary to iodides and bromides, can normally be precipitated from a neutral solution.

In the next step, the precipitate is centrifuged and washed with three quantities, each of 1 ml, of *water R*. This operation should be carried out rapidly in subdued light, disregarding the fact that the supernatant solution may not become perfectly clear. This is necessary since the precipitate when exposed to light turns grayish or black as Ag^+ is reduced to metallic silver.[1]

$$2AgCl\downarrow + \lambda \rightarrow 2Ag\downarrow + Cl_2\uparrow$$

The precipitate is suspended in 2 ml of *water R* and 1.5 ml of *ammonia R* is added. The precipitate dissolves easily with the possible exception of a few large particles, which dissolve slowly — this due to Ag^+ ions' ability to form the diamminoargentate complex.

$$AgCl\downarrow + 2NH_3 \rightarrow [Ag(NHl_3)_2]^- + Cl^-$$

The silver chloride precipitate has to be separated from the original test solution since the ions present here, e.g., the cation of the substance to be examined, can disturb the complexation.[3] Compared to the silver salts of bromide and iodide, silver chloride is the least insoluble. So the formation constant in the above complexation wins in the competition for the silver ion. Silver iodide is more insoluble, and here the precipitate dissolves in ammonia only with difficulty. The silver iodide is not soluble in ammonia at all.

The ions IO_3^- and BrO_3^- can be expected to give similar reactions, although different references disagree somewhat as to what extent, and Br^- is also likely to give a false positive reaction, although silver bromide is pale yellow rather than white. But perhaps the greatest disadvantage of the method is not its relatively high numbers of candidates for false positive reactions but rather its sensitivity toward interference. The high number of insoluble silver salts lowers the ruggedness of the method, since the presence of even low concentrations of organic and inorganic cations gives precipitates of various colors.[2,3]

Test (b)

Into a test tube is introduced a quantity of the substance to be examined equivalent to about 15 mg of chloride (Cl^-) or the prescribed quantity. Add

0.2 g of *potassium dichromate R* and 1 ml of *sulfuric acid R*. In the presence of chloride and sulfuric acid, the dichromate is converted into chromyl chloride.

$$4Cl^- + Cr_2O_7^{2-} + 6H^+ \rightarrow 2CrO_2Cl_2\uparrow + 3H_2O$$

Chromyl chloride is a volatile liquid with a boiling point of 116.5°C, and some of it will escape the solution as gas. At the same time, however, a small amount of chlorine gas can be expected to evolve through the reaction.

$$6Cl^- + Cr_2O_7^{2-} + 14H^+ \rightarrow 3Cl_2\uparrow + 7H_2O$$

In the second step of the analysis, a filter paper strip impregnated with 0.1 ml of *diphenylcarbazide solution R* is placed over the opening of the test tube. Some of the chromyl chloride will be captured by the filter paper impregnated with reagent and will turn violet-red as a consequence of the two consecutive reactions shown here (Figure 3.16.1).

The chromium of chromyl chloride is chromium(VI), which will oxidize diphenylcarbazide to diphenylcarbazone (Figure 3.16.1). The chromium(III) produced in this reaction forms a complex (Figure 3.16.2) with the diphenylcarbazone, which gives the very strong color observed. Several resonance forms of the exemplary structure shown exist.

It is stressed in the text that the impregnated paper must not come into contact with the potassium dichromate. Since potassium dichromate is a

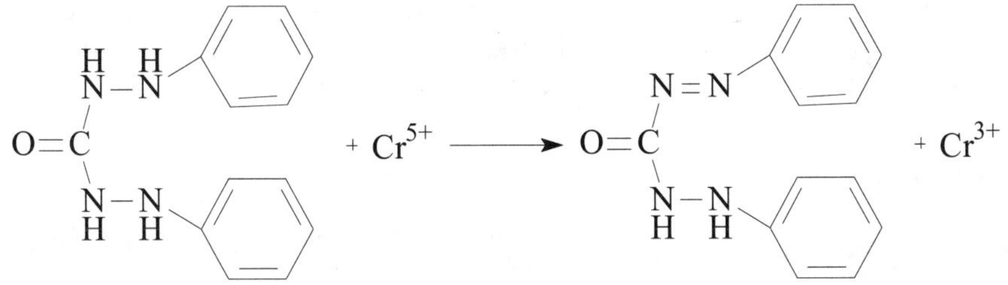

Figure 3.16.1 Oxidation.

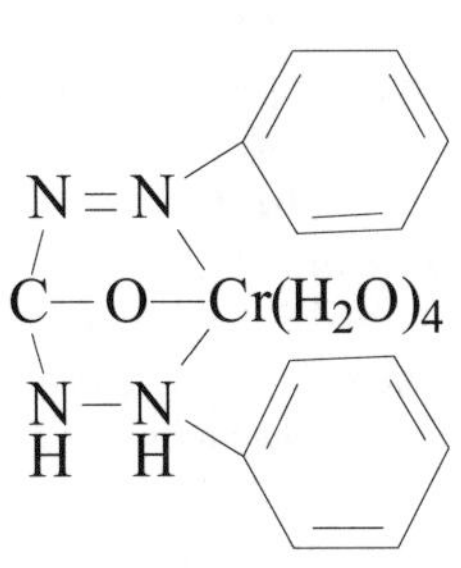

Figure 3.16.2 Diphenylcarbazone complex.

chromium(VI) compound like chromyl chloride, strains of the reagent will make the color reaction evolve. The color reaction is very sensitive and has been used to determine chromium in the ppm area. There has been some debate about whether chromium(III) or chromium(II) is responsible for the reaction and the coordination between the molecules of the complex, but this apparently has not been disclosed.[5]

The ability of chloride to form a chromyl compound is not unique within the halogens. When bromide and iodide are treated with sulfuric acid and dichromate, they are converted into their free halogens. But fluorides will under these conditions form volatile chromyl fluoride, CrO_2F_2, and therefore give a reaction similar to chloride. The mercury chloride does not respond to the test owing to its low solubility, and the formation of chromyl chloride is limited for lead, silver, antimony, and tin chloride for the same reason. Iodide in 1:15 relative to chloride will give free chlorine. The presence of nitrites and nitrates interferes, as nitrosyl chloride is formed.[1]

The color reaction in the strip is quite selective. References dealing with the color reaction used in quantitative work state that interfering ions are iron, vanadium, molybdenum, copper, and mercury. But this is the case only when they are present in very large excess over chloride, so the relevance to the identification test is questionable.[6] If chromium(VI) is present in excess in the strip, the diphenylcarbazone will be further oxidized, to give a yellow appearance, and eventually degraded completely. The reagent has to be prepared fresh, since it oxidizes and darkens quickly when stored.[1,4,7–9]

With the exception of fluoride, this test is said to have a high degree of selectivity, since several conditions have to be met to give a positive reaction. Chromium(VI) has to be brought to a gaseous phase to come into contact with the color reagent. The reagent does not come in contact with the test solution, which minimizes the risk of interference.

References

1. Vogel, A.I. and Svehla, G., *Vogel's Qualitative Inorganic Analysis*, 6th ed., Longman Scientific & Technical, Essex, 1987, pp. 57, 63, 66, 76, 174, 246–249.
2. Charlot, G., *Qualitative Inorganic Analysis. A New Physio-Chemical Approach*, Methuen, London, p. 259.
3. Hartke, K., 2.3.1. Identitätsreaktionen (10. Lfg. 1999), in *Arzneibuch-Kommentar zum Europäische Arzneibuch*, Band I, *Allgemeiner Teil*, Hartke, K. et al., Wissenschaftliche Verlagsgellschaft, Stuttgart/Govi-Verlag — Pharmazeutischer Verlag, Escbor.
4. Kolthoff, I.M. and Elving, P.J., *Treatise on Analytical Chemistry*, Part II, Vol. 7, Interscience, New York, 1961, p. 361.
5. Marchart, H., Über die reaction von chrom mit diphenylcarbazid und diphenylcarbazon, *Analytica Chimica Acta*, 30, 11, 1964.
6. Fries, J. and Getrost, H., *Organische Reagenzien für die Spurenanalyse*, E. Merck, Darmstadt, 1975, p. 114.
7. Reimers, F., *The Basic Principles for Pharmacopoeial Tests*, Heinemann Medical Books, London, 1956, p. 21.

8. Kolthoff, I.M. and Elving, P.J., *Treatise on Analytical Chemistry*, Part II, Vol. 8, Interscience, New York, 1963, pp. 313, 316, 338.
9. Lange, B. and Vejdĕlek, Z.J., *Photometrische Analyse*, Verlag Chemie, Wienheim, 1980, p. 88.

3.17   Citrates (European Pharmacopoeia 2.3.1)

The test identifies the substance to be examined as a salt of citric acid (Figure 3.17.1). Reference to citrates is made in less than 10 monographs, and they are about evenly distributed between simple alkali metal salts of citric acid and citrate salts of cationic active pharmaceutical ingredients. Lithium citrate can be considered a member of the latter group, but the other inorganic salts are used as anticoagulants and as buffers. Citric acid itself is a diuretic and raises the pH of the urine. A special case is triethyl citrate, which is the triethyl ester of citric acid. Citric acid is a crystalline solid with a high water solubility. Alkali metal salts are water soluble, but other metal salts are sparingly soluble. Citric acid has three carboxylic acid groups with pKa values of 3.1, 4.8, and 6.4, respectively. Citric acid is widely distributed in plants and in animal tissues and fluids, but the usual source is industrial fermentation.

A color reaction based on the inorganic reagent disodium pentacyano nitrosylferrate $Na_2[FE(CN)_5(NO)]$, commonly called sodium nitroprusside, is used in the test. This reagent has been widely used for two different reactions that have formed the basis of photometric determinations of a large variety of compounds.

In the Legal reaction, discovered in 1883 by E. Legal, nitroprusside reacts with the α-carbon of enolizable ketones and aldehydes. The exact mechanism is not established, but it is clear that the first step involves the enolate ion of the substance to be examined. A compound having an unsaturated α-carbon will in solution partly be present as the resonans form called the enol form as opposed to the parent keto form. The ketol–enol equilibrium using acetone as an example is shown here (Figure 3.17.2).

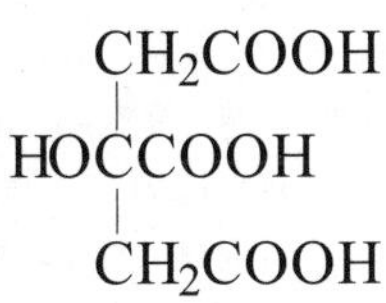

$$CH_2COOH$$
$$|$$
$$HOCCOOH$$
$$|$$
$$CH_2COOH$$

Figure 3.17.1 Citric acid.

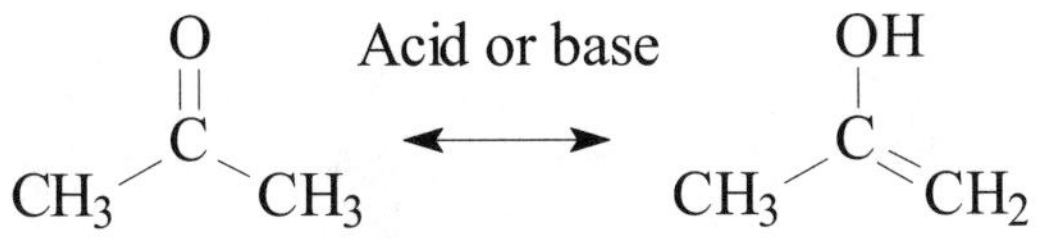

Figure 3.17.2 Ketol-enol equilibrium.

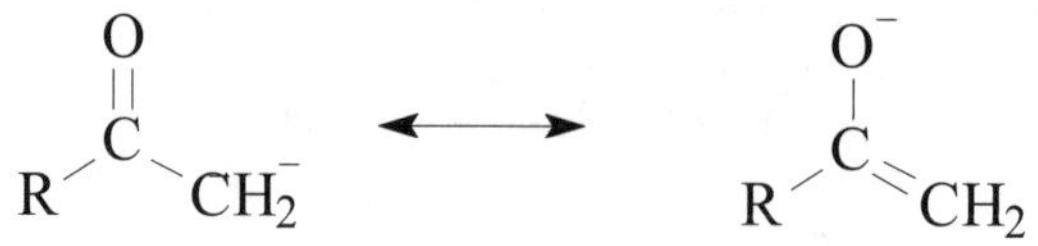

Figure 3.17.3 Resonance.

Both an acidic and an alkaline environment will catalyze a shift of the equilibrium to the right, making the portion of the molecule actually present in the enol form higher. This equilibrium runs through a number of steps that are different depending on whether the solution is acid or alkaline. If the solution is alkaline, the keto–enol interconversion occurs via the enolate ion. In this enolate ion a proton is abstracted from either the alcohol group or the α-carbon of the enol form, and the two forms are stabilized by resonance (Figure 3.17.3).

Overall, this equilibrium gives the α-carbon an acidic character, since it is to some extent found short of one proton and with a negative charge. Such a carbon is susceptible to reaction by other organic compounds and also inorganic compounds. The sodium nitroprusside reagent reacts with the acid α-carbon and yields, through a series if intermediates, the product given below. Again acetone is given as an example in Figure 3.17.4.

Other structures (Figure 3.17.4) containing more than one acetone entity have been proposed, however. The product of the reaction is normally red to orange-red or blue, depending on the pH of the solution. Not much specificity can be claimed, since the test can be used for the determination of, in principle, any substance containing an acidic α-carbon. This is fulfilled by many different aldehydes and ketones but also alkaloids and other nitrogen- or even sulfur-containing compounds.

In the Rimini reaction, discovered in 1898 by E. Rimini, nitroprusside reacts with secondary amines in the presence of an aldehyde or ketone, to give a red to orange-red or blue reaction product. The reaction has been used in the determination both of amines and of aldehydes/ketones. Most of the determinations published of acetaldehyde with nitroprusside use this variant of the reaction. There has been some discussion about the reaction mechanism.

The Legal reaction forms the basis of the citrate identification test, and exactly the same methodology is used in the test 3.21. Lactates. In the first part of analysis the citric acid is oxidized to a substance that is susceptible to the nitroprusside reagent.

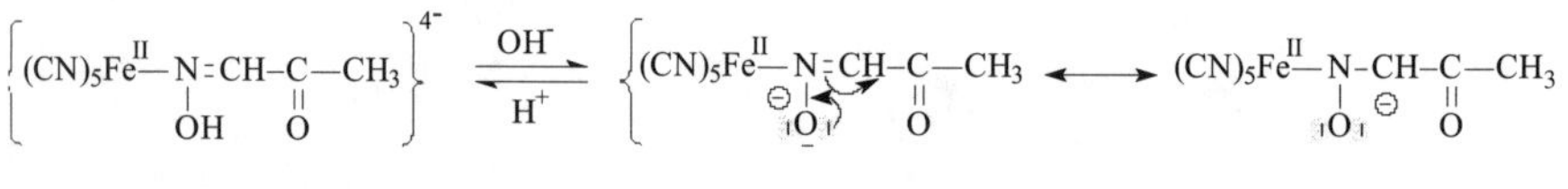

Figure 3.17.4 Reaction product.

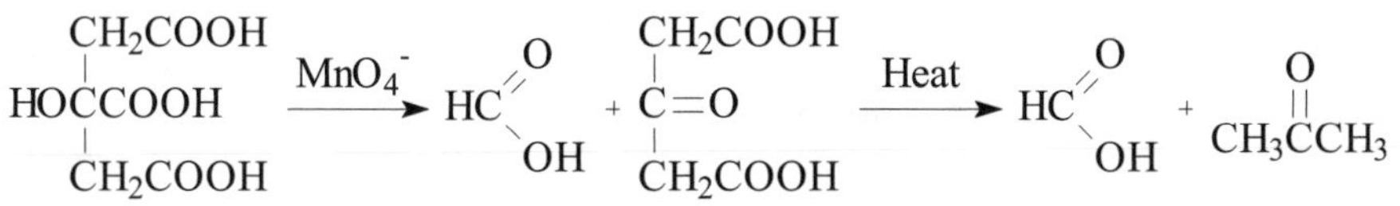

Figure 3.17.5 Oxidation and decarboxylation.

A test solution is prepared by dissolving a quantity of the substance to be examined equivalent to about 50 mg of citric acid in 5 ml of *water R*, or 5 ml of the prescribed solution is used. Add 0.5 ml of *sulfuric acid R* and 1 ml of *potassium permanganate solution R*. The test solution is warmed until the color of the permanganate is discharged.

When citric acid is exposed to permanganate in sulfuric acid, it is oxidized to formic acid and 1,3-dicarboxylic acid acetone, and the latter is changed, through decarboxylation caused by heat, into formic acid and acetone (Figure 3.17.5).

Add 0.5 ml of a 100 g/l solution of *sodium nitroprusside R* in dilute *sulfuric acid R* and 4 g of *sulfamic acid R*. The solution is made alkaline with *concentrated ammonia R*, added dropwise until all the sulfamic acid has dissolved. Addition of an excess of *concentrated ammonia R* produces a violet color, turning to violet-blue. The product formed is the one used for example above (Figure 3.17.4). The sulfamic acid is added with the purpose of binding any nitrous gases formed.[2–8]

References

1. Roth, H.J. and Surborg, K.H., Zum mechanismus und sur spezifität der legalschen probe, *Archiv der Pharmazie*, 301, 686, 1968.
2. Veibel, S., *The Identification of Organic Compounds*, 7th ed. (4th English ed.), G.E.C. GAD, Copenhagen, 1971, pp. 130, 391, 407.
3. Kakáč, B. and Vejdělek, Z.J., *Handbuch der Photometrische Analyse Organischer Verbindungen*, Band 1, Verlag Chemie, Weinheim, 1974, pp. 179, 201, 202, 285, 487, 537.
4. Kakáč, B. and Vejdělek, Z.J., *Handbuch der Photometrische Analyse Organischer Verbindungen*, Band 2, Verlag Chemie, Weinheim, 1974, pp. 734, 846,947, 1033, 1101.
5. Kakáč, B. and Vejdělek, Z.J., *Handbuch der Photometrische Analyse Organischer Verbindungen*, 1. Ergänzungsband, Verlag Chemie, Weinheim, 1977, pp. 29, 66, 82, 149, 215, 312.
6. Kakáč, B. and Vejdělek, Z.J., *Handbuch der Photometrische Analyse Organischer Verbindungen*, 2. Ergänzungsband, Verlag Chemie, Weinheim, 1983, pp. 59, 76, 328.
7. Vejdělek, Z.J. and Kakáč, B., *Farbreaktionen in der Spectrophotometrischen Analyse Organischer Verbindungen*, Band II, VEB Gustav Fischer Verlag, Jena, 1973, p. 280.
8. Pesez, M. and Bartos, J., *Colometric and Fluorimetric Analysis of Organic Compounds and Drugs*, Marcel Dekker, New York, 1974, pp. 144, 157, 198, 276, 385, 493.

3.18   Esters (European Pharmacopoeia 2.3.1)

The test identifies that the substance to be examined contains an ester group (Figure 3.18.1). Since in principle an infinite number of organic compounds contain an ester group, even a true positive reaction does not afford proof about which substance is really at hand. Perhaps for this reason reference to ester is given in only three monographs, despite the hundreds of substances in the pharmacopoeia containing one or more ester groups.

The ester group is regarded as the product of a condensation of a carboxylic acid and an alcohol. In acidic or alkaline solution this reaction is reversed, and the ester is hydrolyzed into the parent carboxylic acid and alcohol. The sensibility toward hydrolysis depends upon the R substituents. Esters are in most cases insoluble in water, unless the molecule contains other hydrophilic groups.

The test is based on a reaction with hydroxylamine and iron(III), which formerly found very widespread application.

In the first step of the analysis, up to about 30 mg of the substance to be examined or the prescribed quantity are added 0.5 ml of a 70 g/l solution of *hydroxylamine hydrochloride R* in *methanol R* and 0.5 ml of a 100 g/l solution of *potassium hydroxide R* in *alcohol R*. The solution is heated to boiling. This procedure liberates the alcohol (Figure 3.18.2) and converts the carboxylic acid into a hydroxamic acid (Figure 3.18.3).

The conditions (pH and time allowed for reaction) necessary to complete the reaction depend, as mentioned above, upon the actual ester to be determined and its substituents.

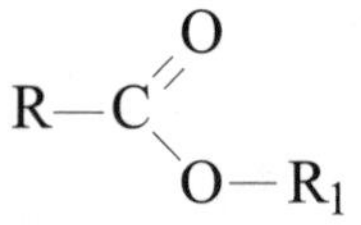

Figure 3.18.1 The ester functional group.

Figure 3.18.2 Ester hydrolysis.

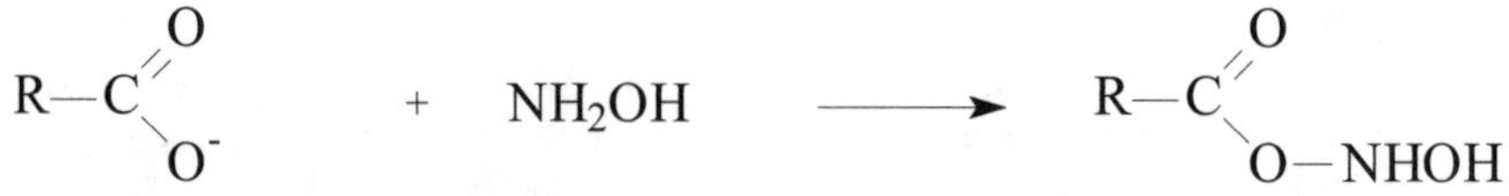

Figure 3.18.3 Hydroxamic acid formation.

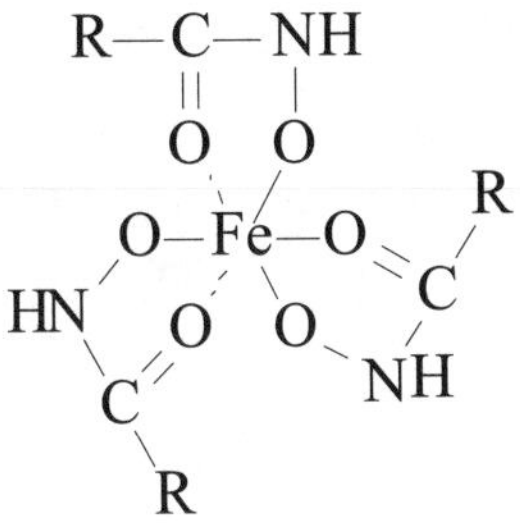

Figure 3.18.4 Iron (III) complex.

In the next step, the test solution is cooled and acidified with dilute *hydrochloric acid R*. Add 0.2 ml of *ferric chloride solution R1* diluted ten times and a bluish-red or red color is produced. The product responsible for the color is a complex of iron(III) hydroxamates (Figure 3.18.4).

The stability of the iron complex depends upon the test solution pH and the iron(III) concentration.[1]

The reaction is neither specific nor selective, since many carboxylic acids, acid anhydrides, acid halides, acid amides and imides, trihalo compounds, and even lactones give the same reaction.[2-4] Also many nitrogen-bearing compounds give a positive reaction. Some references suggest that a blank be prepared in which the substance to be examined is treated as described, but without converting it into a hydroxamic acid, because some substances give a color reaction alone.[5]

A large number of articles have been published describing methods for determinations of these compounds using the same methodology. A few compounds give a yellow product, mainly esters of carbonic acid, carbamic acid, sulphonic acid, and inorganic acids.[6]

References

1. Pesez, M. and Bartos, J., *Colometric and Fluorimetric Analysis of Organic Compounds and Drugs*, Marcel Dekker, New York, 1974, p. 316.
2. Kakáč, B. and Vejdělek, Z.J., *Handbuch der Photometrische Analyse Organischer Verbindungen*, Band 1, Verlag Chemie, Weinheim, 1974, pp. 361, 364, 371, 382.
3. Kakáč, B. and Vejdělek, Z.J., *Handbuch der Photometrische Analyse Organischer Verbindungen*, 1. Ergänzungsband, Verlag Chemie, Weinheim, 1977, p. 144.
4. Kakáč, B. and Vejdělek, Z.J., *Handbuch der Photometrische Analyse Organischer Verbindungen*, 2. Ergänzungsband, Verlag Chemie, Weinheim, 1983, pp. 153, 155.
5. Vogel, A.I., *Elementary Practical Organic Chemistry*, Part 2, *Qualitative Organic Analysis*, 5th ed., Longmans, London, 1962, p. 506.
6. Veibel, S., *The Identification of Organic Compounds*, 7th ed. (4th English ed.), G.E.C. GAD, Copenhagen, 1971, p. 236.

3.19 Iodides (European Pharmacopoeia 2.3.1)

The test identifies the substance to be examined as an iodide (I) salt or a substance containing chemically bound iodine that can be released as iodide upon treatment. At the present only three monographs make reference to iodides, two being simple alkali metal iodide salts and the third a triiodide of a nitrogen base. Contrary to chloride, the physiological role of iodide discourages its use as an inactive counter-ion. Most simple inorganic salts of iodide are water soluble; exceptions are the salts of lead, thallium, silver, mercury, and copper. In general, however, the iodide salts are less soluble than the corresponding chloride and bromide salts. Iodine in itself is rather water insoluble, but the solubility is enhanced by the presence of iodide through the formation of a soluble anionic complex.

$$I_2 + I^- \rightarrow I_3^-$$

The halogens are typically identified through precipitation with silver, and this precipitation is utilized in test (a). The solubility and color of iodine in chloroform are the basis of test (b).

Test (a)

The test is identical to test (a) of 3.16. Chlorides and test (a) of 3.13. Bromides, and the three anions are distinguished partly by the color of the silver precipitate formed, but primarily by the respective precipitates' solubility in ammonia. However, since the silver ion produces insoluble salts with a large number of organic and inorganic cations, there is a substantial risk of both a false positive reaction and of interference. The description below, therefore, is not exclusive and should be read together with 3.16. Chlorides and 3.13. Bromides.

A test solution is prepared by dissolving a quantity of the substance to be examined equivalent to about 4 mg of iodide (I⁻) in 2 ml of *water R* or by using 2 ml of the prescribed solution. It is acidified with dilute *nitric acid R*, and 0.4 ml of *silver nitrate solution R1* is added. Shake and allow to stand. A curdled, pale-yellow precipitate of silver iodide is formed.

$$Ag^+ + I^- \rightarrow AgI\downarrow$$

Since the precipitation is done from a test solution containing nitric acid, a large number of substances that will only precipitate under neutral conditions can be excluded from being responsible for the precipitate formed. For example, a number of carboxylic acids give silver precipitates that are soluble in dilute nitric acid. But still a large number of cations cannot be excluded at this point, and it is even difficult to differentiate between the iodide, chloride, and bromide precipitates, since their color differences are not always very marked. Silver chloride is almost white, whereas silver iodide and silver bromide appear yellow or pale yellow.

In the next step, the precipitate is centrifuged and washed with three quantities, each of 1 ml, of *water R*. Carry out this operation rapidly in subdued light, disregarding that the supernatant solution may not become perfectly clear. This precaution is necessary since the precipitate, when exposed to light, turns grayish or black as Ag^+ is reduced to the metallic silver.[1]

$$2AgI\downarrow + \lambda \rightarrow 2Ag\downarrow + I_2$$

The precipitate is suspended in 2 ml of *water R*, and 1.5 ml of *ammonia R* is added. The precipitate must not dissolve. Silver iodide, compared to silver chloride and bromide, is relatively insoluble. So, where these two salts would dissolve because the silver ion forms a complex with ammonia, is not the case with silver iodide. The reason for isolating the precipitate is to avoid the ions of the original test solution, which could disturb the complex formation.

As with chloride and bromide, there is no total agreement in references on which ions are capable of giving false positive reactions, but perhaps the greatest disadvantage of the method is not its relatively high number of candidates for false positive reactions but rather its sensitivity toward interference. The high number of insoluble silver salts lowers the ruggedness of the method, since the presence of even low concentrations of organic and inorganic cations gives precipitates of various colors.[2]

Test (b)

A test solution of the substance to be examined containing about 5 mg of iodide (I^-) per ml is prepared. To 0.2 ml of this solution, or to 0.2 ml of the prescribed solution, 0.5 ml of dilute *sulfuric acid R*, 0.1 ml of *potassium dichromate solution R*, 2 ml of *water R*, and 2 ml of *chloroform R* are added. Shake for a few seconds and allow to stand. The chloroform layer is colored violet or violet-red. The reason for this appearance is that iodide in the acid environment is oxidized by the dichromate to iodine.

$$6I^- + Cr_2O_7^{2-} + 7H_2SO_4 \rightarrow 3I_2 + 2Cr^{3+} + 7SO_4^{2-} + 7H_2O$$

The iodine produced is more soluble in the (lower) chloroform phase, in which it is easily recognized by the color. One should not be confused by the orange-red appearance of the water phase; this is given by the dichromate ion.

The exact color of iodine in an organic phase depends on the solvent used and the purity of the solvent. In chlorinated solvents, it is violet or violet-red, but brown in alcohol, ethers, and ketones. In pure chloroform the color is violet, but if it is contaminated with alcohol, this will affect the color to be more violet-red. Bromide will react similarly to the conditions of the test, but the color of Br_2 in chloroform is red-brown.[2-4]

If an excess of a stronger oxidizing substance is used instead of dichromate, there is a risk that iodide will be further oxidized to iodate, which is colorless and insoluble in chloroform.[5]

References

1. Vogel, A.I. and Svehla, G., *Vogel's Qualitative Inorganic Analysis*, 6th ed., Longman Scientific & Technical, Essex, 1987, pp. 57, 63, 66, 76, 178, 246–249.
2. Charlot, G., *Qualitative Inorganic Analysis. A New Physio-Chemical Approach*, Methuen, London, 1954, p. 259.
3. Hartke, K., 2.3.1. Identitätsreaktionen (10. Lfg. 1999), in *Arzneibuch-Kommentar zum Europäische Arzneibuch*, Band I, *Allgemeiner Teil*, Hartke, K. et al., Wissenschaftliche Verlagsgesellschaft, Stuttgart/Govi-Verlag — Pharmazeutischer Verlag, Escbor.
4. Reimers, F., *The Basic Principles for Pharmacopoeial Tests*, Heinemann Medical Books, London, 1956, p. 21.
5. Kolthoff, I.M. and Elving, P.J., *Treatise on Analytical Chemistry*, Part II, Vol. 7, Interscience, New York, 1961, p. 361.

3.20   Iron (European Pharmacopoeia 2.3.1)

The test identifies that the substance to be examined contains iron. Reference to iron is made in five monographs, and they include metallic iron itself and simple organic and inorganic salts of Fe^{2+} and Fe^{3+}. These substances are widely used in vitamin supplements. Four monographs dictate test (a) and one dictates test (c). Being a typical transition metal, iron can exist in more that one oxidation level, but out of these only iron(II) and iron(III) are stable, and they are both very common. Both ions form insoluble hydroxides (white and reddish-brown, respectively), but iron(II) is very easily oxidized into iron(III) at neutral and alkaline conditions. Even dissolved oxygen affords this reaction, so solutions of iron(II) have to be acidic to be stable. A solution containing iron(II) gives a green appearance, whereas solutions of iron(III) are pale yellow.

Since a main objective of the test is to differentiate between Fe^{2+} and Fe^{3+}, and since this is obtained by using Fe^{2+} and Fe^{3+} reagents, it is important to recapitulate how these compounds are named. Iron in oxidation level two, Fe^{2+}, is called iron(II), and its salts are called ferro compounds (e.g., potassium ferrocyanide) or ferrate(II) compounds (hexacyanoferrate(II)). Iron in oxidation level three, Fe^{3+}, is called iron(III), and its salts are called ferri compounds (e.g., potassium ferricyanide) or ferrate(III) compounds (hexacyanoferrate(III)).

To be able to identify both iron(II) and iron(III) and distinguish between them, three different identification tests are described in the pharmacopoeia.

Test (a)

The test identifies iron(II) and differentiates it from iron(III).

Fe^{2+}, Fe^{3+}, and CN^- will under appropriate conditions form a blue complex or even a precipitate of $KFe[Fe(CN)_6]$ called prussian blue. If only Fe^{2+} or only Fe^{3+} is present, the solution will appear yellow due to the formations of either of the two complexes shown here.

$$Fe^{2+} + 6CN^- \rightarrow [Fe(CN)_6]^{4-}$$

$$Fe^{3+} + 6CN^- \rightarrow [Fe(CN)_6]^{3-}$$

These complexes are so stable that the iron can be liberated only by radical procedures such as boiling the solution with concentrated sulphuric acid, and their potassium salts are then easily isolated. These salts are used as reagents, so that the potassium hexacyanoferrate(III) (potassium ferricyanide) is used to identify iron(II), and potassium hexacyanoferrate(II) (potassium ferrocyanide) is used to identify iron(III).

$$Fe^{2+} + K^+ + [Fe(CN)_6]^{3-} \rightarrow KFe[Fe(CN)_6]$$

$$Fe^{3+} + K^+ + [Fe(CN)_6]^{4-} \rightarrow KFe[Fe(CN)_6]$$

The complex formed is in both cases prussian blue, although earlier literature suggests that the product of the lower reaction was of a stoichiometrically different composition named Turmbull's blue. This has later been disproved in several articles. It should also be recognized that the above formula for prussia blue, $KFe[Fe(CN)_6]$, is not intended to give an absolute statement of the stochiometry of the complex, and some authors use the formula $Fe_4[Fe(CN)_6]_3$. The composition and three-dimensional structure is, however, more complicated than suggested by these simple formulas, as is shown in several references.[1]

A test solution is prepared by dissolving a quantity of the substance to be examined equivalent to about 10 mg of iron (Fe^{2+}) in 1 ml of *water R* or by using 1 ml of the prescribed solution. Add 1 ml of *potassium ferricyanide solution R* and a blue precipitate of prussian blue is formed that does not dissolve on addition of 5 ml of dilute *hydrochloric acid R*.

If the iron present is in fact 100% in oxidation state II, a white precipitate of iron(II) hexocyanoferrate(II) is formed (no structure or formula is attempted). This will take place only in the absence of air and other oxidizing substances, since sufficient amounts of iron(III) would rapidly be formed to allow the formation of prussian blue. If an oxidizing substance is present, a small amount of iron(III) will be formed, giving prussian blue. If the test is performed under alkaline conditions, white $Fe(OH)_2$ might precipitate, which subsequently is oxidized to reddish-brown, $Fe(OH)_3$. If the reaction is performed without adding dilute hydrochloric acid, prussian blue may not precipitate but stay in the solution.

The test is stated to be very selective, but a few interferences are possible. Redox reagents interfere by taking all iron to either oxidation level two or three. Ions that form complexes with iron, e.g., fluoride or oxalates, can ultimately prevent the prussian blue reaction.[2]

Test (b)

The test identifies iron(III) and differentiates it from iron(II).

A quantity of the substance to be examined, equivalent to about 1 mg of iron (Fe^{3+}), is dissolved in 30 ml of *water R*. To 3 ml of this solution or to 3 ml of the prescribed solution, 1 ml of dilute *hydrochloric acid R* and 1 ml of *potassium thiocyanate solution R* are added. The solution is colored red by an iron(III) thiocyanate complex.

$$Fe^{3+} + 3SCN^- \rightarrow Fe(SCN)_3$$

Iron(II) does not form a corresponding $Fe(SCN)_2$ and gives no reaction toward this test. It should be remembered that iron(II) solutions often contain at least some iron(III), especially if the solution is not acidified, and this impurity would give a positive reaction.

In the second step of analysis, take two portions of the mixture, each of 1 ml. To one portion add 5 ml of *isoamyl alcohol R* or 5 ml of *ether R*. Shake and allow to stand. The organic layer is colored pink.

Since the iron(III) thiocyanate complex is neutral, it is more soluble in the organic phases than in water.

To the other portion add 2 ml of *mercuric chloride solution R*. The red color disappears because mercury(II) forms a stronger, but colorless, complex than iron(III) with the thiocyanate ion.

$$2Fe(SCN)_3 + 3Hg^{2+} \rightarrow 2Fe^{3+} + 3Hg(SCN)_2$$

Colored salts like copper, chromium, cobalt, and nickel will reduce the sensitivity of the test, and all heavy metals are expected to interfere. No elements are stated to give a false positive reaction, but a number of ions can interfere. Reductants, like tin(II), can reduce Fe^{3+} to Fe^{2+} and will (ultimately) give a false negative result. Ions capable of forming strong complexes with the ions of the test are another cause of interference. Fluoride, acetate, oxalate, and tartrate are examples.

$$Fe(SCN)_3 + 6F^- \rightarrow [FeF_6]^{3-} + 3SCN^-$$

These complexes are all ionic and will not dissolve in the organic layer. Also nitrites interfere by forming NOSCN that gives a color similar to $Fe(SCN)_3$.[2]

Some references describe the method applied for colorimetric determination of iron(II), and one states the number of species causing interference as many.[3,4]

Test (c)

The test identifies iron(III) and differentiates it from iron(II). The chemistry involved in the test is identical to what is described in test (a) and is dealt with in more detail above.

A test solution is prepared by dissolving a quantity of the substance to be examined equivalent to not less than 1 mg of iron (Fe^{3+}) in 1 ml of *water R* or by using 1 ml of the prescribed solution. Add 1 ml of *potassium ferrocyanide solution R* and a blue precipitate, prussian blue, is formed that does not dissolve on addition of 5 ml of dilute *hydrochloric acid R*.

If the iron present is in fact 100% in oxidation level three, a brown coloration or precipitate is produced owing to a complex containing iron(III) heaxcyanoferrate(III) (no stucture or formula is attempted). If a reducing substance is present, some iron(III) is reduced to iron(II), and prussian blue will be formed.

References

1. Keggin, J.F. and Miles, F.D., Structure and formulæ of the prussian blues and related compounds, *Nature*, 58, 577, 1936.
2. Feigl, F., *Qualitative Analysis by Spot Tests*, 3rd ed., Elsevier, New York, 1947, p. 124.
3. Allport, N.L., *Colorimetric Analysis*, Chapman & Hall, London, 1945, p. 53.
4. Kodama, K., *Methods of Quantitative Inorganic Analysis*, Interscience, New York, 1963, p. 257.

3.21 Lactates (European Pharmacopoeia 2.3.1)

The test identifies the substance to be examined as a salt of lactic acid (Figure 3.21.1). At the present seven monographs reference lactates, and besides lactic acid itself, the majority of monographs are of various inorganic salts of the acid. Lactate has a potential use as an indifferent counter-ion for organic cationic active pharmaceutical ingredients, but only a single monograph, ethacridine lactate monohydrate, describes one such. Lactic acid is an enatiomeric substance, but milk, which is the most common source of the

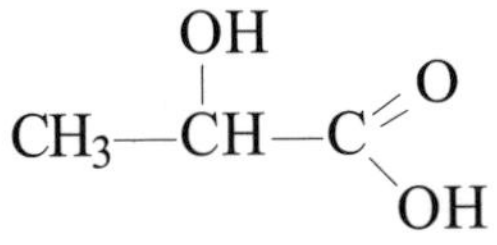

$$CH_3-\underset{\underset{\displaystyle OH}{|}}{CH}-C\overset{\displaystyle O}{\underset{\displaystyle OH}{\diagup\diagdown}}$$

Figure 3.21.1 Lactic acid.

substance, contains the racemate. Technically it can be prepared by fermentation of various carbohydrates, among others glucose. It has one carboxylic
acid group with a pKa value of 3.8. It is soluble in water, alcohol, and furfurol
but less soluble in ether. It is practically insoluble in chloroform, petroleum
spirits, and carbon disulphide. The acid form is a crystalline powder that
has a boiling point of 16.8°C, and the salts are likewise usually crystalline
substances.

A color reaction based on the inorganic reagent disodium pentacyano
nitrosylferrate $Na_2[FE(CN)_5(NO)]$, most commonly called sodium nitroprusside, is used in the test. This reagent has been widely used for two different
reactions that have formed the basis of photometric determinations of a large
variety of compounds.

In the Legal reaction, discovered in 1883 by E. Legal, nitroprusside reacts
with the α-carbon of enolizable ketones and aldehydes. The exact mechanism
is not established, but it is clear that the first step involves the enolate ion
of the substance to be examined. A compound having an unsaturated
α-carbon will in solution partly be present as the resonans form the enol
form as opposed to the parent keto form. The ketol–enol equilibrium using
acetone as an example is shown here (Figure 3.21.2).

Both an acidic and an alkaline environment will catalyze a shift of the
equilibrium to the right, making the portion of molecule actually present in
the enol form higher. This equilibrium runs through a number of steps that
are different depending on whether the solution is acid or alkaline. If the
solution is alkaline, the keto–enol interconversion occurs via the enolate ion.
In this enolate ion a proton is abstracted from either the alcohol group or
the α-carbon of the enol form, and the two forms are stabilized by resonance
(Figure 3.21.3).

Overall, this equilibrium gives the α-carbon an acidic character, since it
is to some extent found short of one proton and with a negative charge. Such
a carbon is susceptible to reaction by other organic compounds but also
inorganic compounds. The sodium nitroprusside reagent reacts with the acid

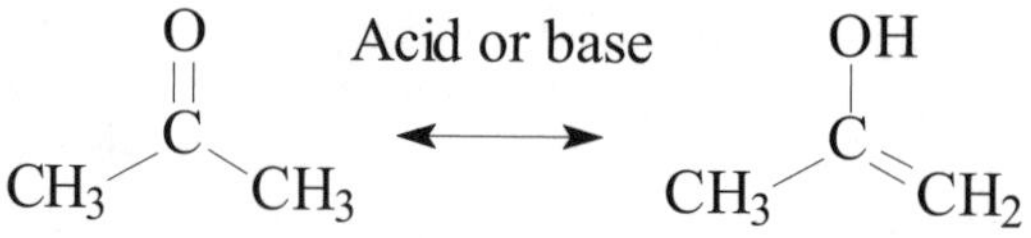

Figure 3.21.2 Ketol-enol equilibrium.

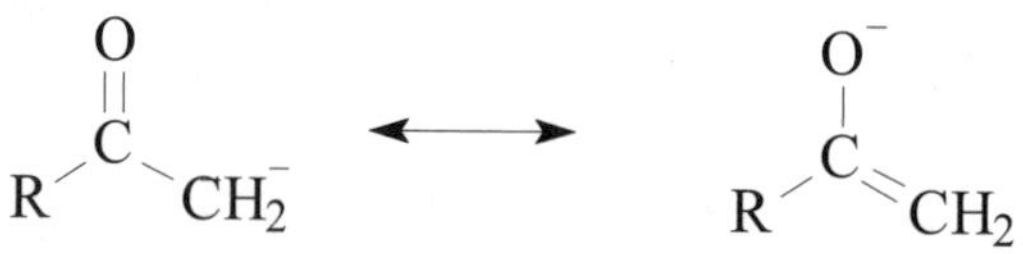

Figure 3.21.3 Resonance.

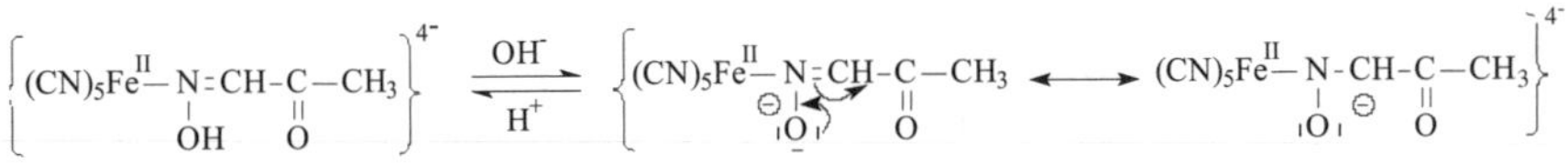

Figure 3.21.4 Reaction product.

α-carbon and yields, through a series if intermediates, the product given below. Again acetone is given as an example (Figure 3.21.4).

Other structures than (Figure 3.21.4), containing more than one acetone entity, have been proposed, however. The product of the reaction is normally red to orange-red or blue, depending on the pH of the solution. Not much specificity can be claimed, since the test can be used for the determination of, in principle, any substance containing an acidic α-carbon. This is fulfilled by many different aldehydes, ketones, but also alkaloids and other nitrogen- or even sulfur-containing compounds.[1]

In the Rimini reaction, discovered in 1898 by E. Rimini, nitroprusside reacts with secondary amines in the presence of an aldehyde or ketone to give a red to orange-red or blue reaction product. The reaction has been used in the determination of both amines and aldehydes/ketones. Most of the determinations published of acetaldehyde with nitroprusside are using this variant of the reaction. There has been some discussion about the reaction mechanism.

The legal reaction forms the basis of the lactates identification test, and exactly the same methodology is used in the test in 3.17. Citrates. In the first part of the analysis the lactic acid is oxidized to a substance which is susceptible to the nitroprusside reagent.

A test solution is prepared by dissolving a quantity of the substance to be examined equivalent to about 5 mg of lactic acid in 5 ml of *water R*, or by using 5 ml of the prescribed solution. Add 1 ml of *bromine water R* and 0.5 ml of *dilute sulphuric acid R*. Heat is applied on a water-bath until the color is discharged, with occasionally stirring with a glass rod.

The bromine added oxidizes the lactic acid to an intermediate, which upon the action of heat loses carbon dioxide by hydrolysis (Figure 3.21.5). The result is formic acid and acetaldehyde.

Add 4 g of *ammonium sulphate R* and the test solution is mixed. Dropwise and without mixing 0.2 ml of a 100 g/l solution of *sodium nitroprusside R* in dilute *sulphuric acid R* is added. Still without mixing, 1 ml of *concentrated ammonia R* is added. After at the most 30 min, a dark green ring appears at

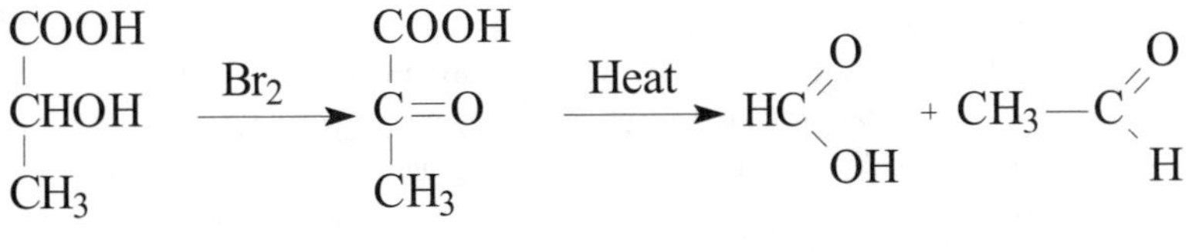

Figure 3.21.5 Oxidation and carbon dioxide loss.

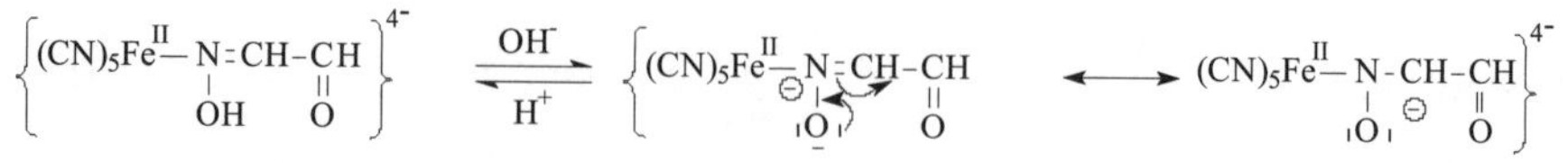

$$\left\{(CN)_5Fe^{II}\!\!-\!\!N\!=\!CH\!-\!CH \atop \quad\;\; \overset{|}{OH}\;\;\; \overset{\|}{O}\right\}^{4-} \;\xrightarrow[H^+]{OH^-}\; \left\{(CN)_5Fe^{II}\!\!\underset{\ominus}{-}\!\!N\!=\!CH\!-\!CH \atop \qquad\quad {}_{\backslash O \,/}\;\;\;\overset{\|}{O}\right\} \;\longleftrightarrow\; \left\{(CN)_5Fe^{II}\!\!-\!\!N\!-\!CH\!-\!CH \atop \qquad\quad {}_{\backslash O \,|}\;\overset{\ominus}{}\;\;\overset{\|}{O}\right\}^{4-}$$

Figure 3.21.6 Reaction product.

the junction of the two liquids. The product (Figure 3.21.6) of the reaction is analogous to the one for acetone shown above.

The test must be performed under alkaline conditions to favor the acetaldehyde enolat ion, and ammonium sulfate and ammonia are selected since they give an intense and more stable coloration.[2-7]

References

1. Roth, H.J. and Surborg, K.H., Zum mechanismus und sur spezifität der legalschen probe, *Archiv der pharmazie*, 301, 686, 1968.
2. Vejdělek, Z.J. and Kakáč, B., *Farbreaktionen in der spectrophotometrischen analyse organischer verbindungen*, Band II, VEB Gustav Fischer Verlag, Jena, 1973, p. 280.
3. Kakáč, B. and Vejdělek, Z.J., *Handbuch der photometrishe analyse organisher verbindungen*, Band 1, Verlag Chemie, Weinheim, 1974, pp. 179, 201, 285, 487, 537.
4. Kakáč, B. and Vejdělek, Z.J., *Handbuch der photometrishe analyse organisher verbindungen*, Band 2, Verlag Chemie, Weinheim, 1974, pp. 734, 846, 947, 1033, 1101.
5. Kakáč, B. and Vejdělek, Z.J., *Handbuch der photometrishe analyse organisher verbindungen*, 1. Ergänzungsband, Verlag Chemie, Weinheim, 1977, pp. 29, 66, 82, 149, 215, 312.
6. Kakáč, B. and Vejdělek, Z.J., *Handbuch der photometrishe analyse organisher verbindungen*, 2. Ergänzungsband, Verlag Chemie, Weinheim, 1983, pp. 59, 76, 328.
7. Pesez, M. and Bartos, J., *Colometric and Fluorimetric Analysis of Organic Compounds and Drugs*, Marcel Dekker, New York, 1974, pp. 144, 157, 198, 276, 385, 493.

3.22   Lead (European Pharmacopoeia 2.3.1)

The test identifies the substance to be examined as a salt of lead, Pb^{2+}. Due to their toxicity salts of lead used earlier in therapy have generally been substituted by salts of other cations. Formerly lead salts such as lead acetate and lead carbonate, were employed as astringents, but this is no longer recommended and at the present no monographs reference the lead identification test. Lead is divalent and even though some references state that the tetravalent ion is possible too, this is without practical implications. Most salts of lead are insoluble, but they have a tendency of forming water soluble complexes such as the tetrachloroplumbate(II) ion $[PbCl_4]^{2-}$ or the tetrahydroxoplumbate(II) $[Pb(OH)_4]^{2-}$.

Lead will form a yellow amphoteric insoluble salt with chromate, and these characteristics are used in Test (a). The water solubility of the halide salts of lead varies strongly with temperature, and this is used in test (b).

Test (a)

A test solution is prepared by dissolving 0.1 g of the substance to be examined in 1 ml of *acetic acid R*, or by using 1 ml of the prescribed solution. 2 ml of *potassium chromate solution R* is added and a yellow precipitate of lead and chromate is formed.

$$Pb^{2+} + CrO_4^{2-} \rightarrow PbCrO_4\downarrow$$

The precipitation will take place in neutral and acidic solutions, and the reason for using acetic acid to control the pH is leads ability to form insoluble salts with the anions of the common mineral acids, e.g., $PbCl_2$ and $PbSO_4$. If the solution is too acidic, no precipitation will occur since the chromate reagent is then shifted onto the orange dichromate form with which it is in equilibrium.

$$2CrO_4^{2-} + 2H^+ \rightarrow Cr_2O_7^{2-} + H_2O$$

In the next step of the analysis, the precipitate is redissolved by adding 2 ml of strong sodium hydroxide solution R. This is because the amphoteric lead forms a water soluble complex in strong alkaline solutions.

$$PbCrO_4\downarrow + 4OH^- \rightarrow [Pb(OH)_4]^{2-} + CrO_4^{2-}$$

It should be remembered that lead forms an insoluble hydroxide salt as a response to sodium hydroxide, in cases where the hydroxide reagent is not added in excess.

$$Pb^{2+} + OH^- \rightarrow Pb(OH)_2\downarrow$$

$$Pb(OH)_2\downarrow \rightarrow [Pb(OH)_4]^{2-}$$

So, in general, controlling the test solution pH is the most important factor in the test, and an obvious starting point for an investigation in the case of an unexpected result. Also barium gives a yellow precipitate with chromate, but since this element is not amphoteric it will not redissolve with excess hydroxide.[1,2]

Test (b)

Unless a solution is prescribed in the monograph, a test solution is prepared by dissolving 50 mg of the substance to be examined in 1 ml of *acetic acid R*.

10 ml of *water R* and 0.2 ml of *potassium iodide solution R* are added and this gives a yellow precipitate of lead iodide.

$$Pb^{2+} + 2I^- \rightarrow PbI_2\downarrow$$

This salt is quite soluble at high temperatures and will dissolve in the second step of the analysis where the test solution is heated to boiling. The silver ion also gives a yellow precipitate with iodide, but this is insoluble even in hot water.

When, in the last step of the analysis, the test solution is allowed to cool, the precipitate is reformed as glistering yellow plates of a characteristic appearance. A possible cause of interference would be any anion contamination, since almost all lead salts are insoluble. Contamination with an oxidizing substance would destroy the iodide reagent by converting it into iodine. [1–3]

References

1. Vogel, A.I. and Svehla, G., *Vogel's Qualitative Inorganic Analysis*, 6th ed., Longman Scientific & Technical, Essex, 1987, p. 57.
2. Charlot, G., *Qualitative Inorganic Analysis. A New Physio-Chemical Approach*, Methuen & Co., London, 1954, p. 212.
3. Vogel, A.I. *A Text-Book of Macro and Semimicro Qualitative Inorganic Analysis*, 4th ed., Longmans, Green and Co., London, 1954, p. 208.

3.23   Magnesium (European Pharmacopoeia 2.3.1)

The test identifies the substance to be examined as a salt of magnesium, Mg^{2+}. Reference to magnesium is currently made in about 15 monographs. With a few exceptions they are simple inorganic salts used as inactive excipients or for the treatment of magnesium deficiency. The carbonate, sulfate, and oxide are used as antacids and laxatives. Magnesium is exclusively found as the bivalent cation Mg^{2+}. Its oxide, hydroxide, carbonate, and phosphate salts are insoluble in water, whereas the rest of its common inorganic salts are water soluble. Due to its insoluble hydroxide, its water solubility shows a high dependency on pH and this is utilized in the identification test. Sample preparation in the majority of the monographs is done by simply dissolving the substance to be examined in water or acidified water. In the case of magnesium stearate the stearic acid is first removed from the test solution by liquid-liquid extraction. Magnesium aspartate dihydrate is ignited to incinerate the aspartate moiety, since it forms a complex with magnesium and thereby prevents the precipitation in step one.[1]

In the first step of the analysis, the 2 ml sample solution is added to 1 ml *dilute ammonia R1*. This shifts the pH up and magnesium forms a white gelatinous precipitate of magnesium hydroxide:

$$Mg^{2+} + 2NH_3 + 2H_2O \rightarrow Mg(OH)_2\downarrow + 2NH_4^+$$

This step is not specific since many other cations form insoluble hydroxides as well.

In the second step, 1 ml of *ammonium chloride solution R* is added to the test tube. This lowers the pH and the precipitate dissolves. In classic inorganic separation procedures, Zn^{2+}, Mn^{2+}, Co^{2+}, and Ni^{2+} will react in the same manner. Only the precipitate of zinc is white, and this dissolves readily in excess ammonia as $Zn(NH_3)_4]^2$, a condition that should have been achieved in step one.

In the last step, 1 ml *disodium hydrogen phosphate solution R* is added. As a result of this addition, a crystalline precipitate of magnesium ammonium phosphate is formed.

$$Mg^{2+} + NH_3 + HPO_4^{2-} \rightarrow Mg(NH_4)PO_4\downarrow$$

This salt is stated to precipitate slowly from dilute solutions because of its tendency to form supersaturated solutions. In such cases, cooling and scratsing on the inner wall of the test tube with a glass rod should speed up the process. If the original test solution, is neutral, and ammonium chloride is not added to *disodium hydrogen phosphate solution R*, magnesium might form a white flocculent precipitate of magnesium hydrogen phosphate.[2]

$$Mg^{2+} + HPO_4^{2-} \rightarrow MgHPO_4, 6H_2O \downarrow$$

Besides the risk of false positive reactions due to the presence of interfering ions, a number of factors could disturb the reaction. Any impurity present in glassware and reagents that prevents the intended manipulation of the pH would give unexpected results, as would the presence of ions which form insoluble salts with the reagents in use.

Reference

1. Hartke, K., 2.3.1. Identitätsreaktionen (10.Lfg.1999), in *Arzneibuch-Kommentar zum europäische arzneibuch*, Band I *Allgemeiner teil*, Hartke, K. et al., Wissenschaftliche Verlagsgellschaft, Stuttgart/Govi-Verlag — Pharmazeutischer Verlag, Escbor.
2. Vogel, A.I. and Svehla, G., *Vogel's Qualitative Inorganic Analysis*, 6th ed., Longman Scientific & Technical, Essex, 1987, p. 140.

3.24 Mercury (European Pharmacopoeia 2.3.1)

The test identifies the substance to be examined as either a salt of mercury, Hg_2^{2+} or Hg^{2+} or a substance containing covalently bound mercury. At the present, reference to mercury is made in only two monographs. One is an inorganic salt mercuric chloride, $HgCl_2$, which has antibacterial action through precipitation proteins and thimerosal (Figure 3.24.1), which likewise is an antiseptic but also a preservative in pharmaceutical product. The use

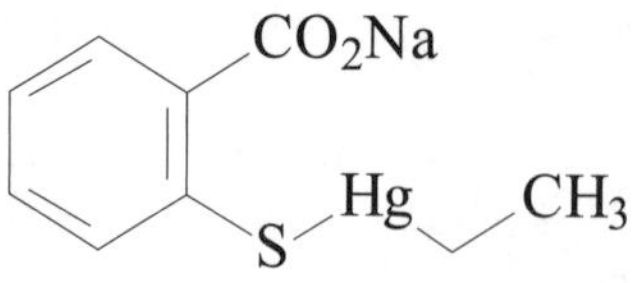

Figure 3.24.1 Thimerosal.

of mercury compound has been diminished due to its toxicological and allergenic properties and the safety of the two above mentioned active ingredients are also questioned.

Mercury can exist in two ionic forms: The mercury(I) ion which is found as the dimer Hg_2^{2+}, and the mercury(II) ion Hg^{2+}. Mercury is unique because it is the only metallic element that is a liquid at room temperature. The chemistry of mercury(I) and mercury(II) is quite different and this allows differentiation using the two tests described. Mercury lies above copper in the electrochemical series, which is used in test (a) as an identification of mercury(II) and mercury(II). Mercury(II) forms an insoluble oxide of a characteristic color and this is used in test (b).

Test (a)

In the first step of analysis, about 0.1 ml of a solution of the substance to be examined is placed on a piece of copper foil. When a solution containing Hg_2^{2+} or Hg^{2+} is brought in contact with metallic copper, a redox reaction takes place.

$$Hg^{2+} + Cu \rightarrow Hg + Cu^{2+}$$

This transfer of electrons takes place spontaneously since the reactions

$$Hg^{2+} + 2e^- \rightarrow Hg_2^{2+}$$

$$Hg_2^{2+} + 2e^- \rightarrow Hg$$

in the electrochemical series lies above the reactions

$$Cu^{2+} + e^- \rightarrow Cu^+$$

$$Cu^+ + e^- \rightarrow Cu$$

As a result of this, metallic mercury will be formed from the ionic mercury ions presents in the test solution and reveal itself as a dark gray stain that becomes shiny on rubbing, using, for example, a dry cloth. Since mercury is volatile, it will evaporate when the foil is heated in a test tube as dictated in the second step of analysis.[1] The foil used has to be well-scraped since copper exposed to the atmosphere will become passivated through its reaction with water, oxygen, and carbon dioxide.[2]

$$2Cu + H_2O + O_2 + CO_2 \rightarrow Cu_2(OH)_2CO_3$$

This forms a protective layer that would prevent the mercury ion from coming in contact with metallic copper.

The demands of the test for a positive reaction, especially the unique feature of mercury's volatility, seem very selective, and hence the identification must be regarded as highly specific. A potential cause of interference could be presence of redox active contaminants in the test solution.

Test (b)

A very large number of cations form insoluble hydroxides and oxides in alkaline solutions, including mercury. The only step of analysis is to add *dilute sodium hydroxide solution R* until it reacts strongly alkaline to litmus paper (above about pH 8.3), a dense yellow precipitate of mercury(II) oxide.

$$Hg^{2+} + 2OH^- \rightarrow HgO_2\downarrow$$

If sodium hydroxide is added in a nonequivalent amount, a brownish-red precipitate of different mercury(I) hydroxides would be formed. Given the same circumstances mercury(I) would first give a black precipitate of Hg_2O.

$$Hg_2^{2+} + 2OH^- \rightarrow Hg_2O\downarrow$$

This could disproportionate into metallic mercury and mercury(II). This precipitate is black due to the metallic mercury.

$$Hg_2O\downarrow \rightarrow Hg\downarrow + HgO\downarrow$$

The test thereby differentiates between mercury(I) and mercury(II).[3,4]

References

1. Reimers, F., *The Basic Principles for Pharmacopoeial Tests*, Heinemann Medical Books, London, 1956, p. 21.
2. Atkins, P.A., *General Chemistry*, Scientific American Books, New York, 1989, p. 777.
3. Vogel, A.I. and Svehla, G., *Vogel's Qualitative Inorganic Analysis*, 6th ed., Longman Scientific & Technical, Essex, 1987, p. 56, 62, 70.
4. Vogel, A.I. *A Text-Book of Macro and Semimicro Qualitative Inorganic Analysis*, 4th ed., Longmans, Green and Co., London, 1954, pp. 212, 218.

3.25   Nitrates (European Pharmacopoeia 2.3.1)

The test identifies the substance examined to be a salt of nitrate, NO_3^-. Nitrate under normal conditions is safe to ingest and this combined with its tendency of forming water-soluble salts makes it a candidate as a salt forming counter-ion for cationic active pharmaceutical ingredients. Of the less than 20 monographs presently referencing nitrates, the majority fall into this category. All inorganic salts of nitrate are water-soluble salts except for the basic salts of mercury and bismuth.

The pharmacopoeia test is quite unique in that a content of nitrate under the conditions given will react with nitrobenzene to give m-dinitrobenzene, which then will react with acetone to form a colored product indication a positive reaction. The methodology of this second reaction was first described in literature in 1939 by Pesez and is based on the Janovsky reaction that for many years was widely used for the colorimetric/photometric determination of substances sometimes referred to as methyl ketones.[1,2]

In the first step of analysis, a mixture of 0.1 ml of *nitrobenzene R* and 0.2 ml of *sulfuric acid R* are added to a quantity of the powdered substance equivalent to about 1 mg of nitrate (NO_3^-) or the prescribed quantity.

When nitrobenzene and nitrate ions are mixed in the presence of sulfuric acid, it reacts by adding nitrate to the aromatic ring (Figure 3.25.1). The main product of this electrofilic substitution is m-dinitrobenzene, but small amounts of the ortho- and para-isomers, (Figures 3.25.2 and 3.25.3) are also produced. M-dinitrobenzene is favored since the NO_2 group is an electron-withdrawing substitute that deactivates the benzene ring but deactivates the meta-position the least. A small quantity of 1,3,5-trinitrobenzene is also produced. The reaction is allowed 5 minutes to complete.

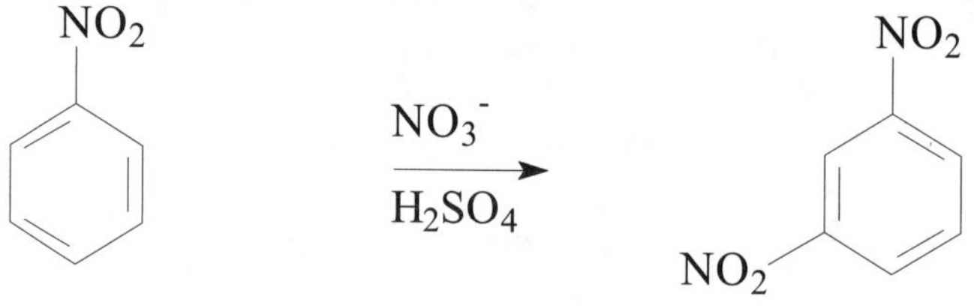

Figure 3.25.1 Addition.

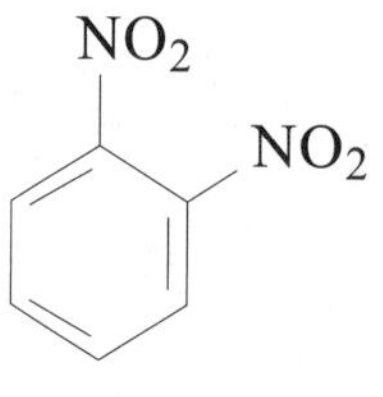

Figure 3.25.2 o-dinitrobenzene.

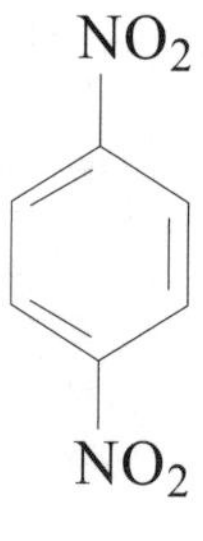

Figure 3.25.3 p-dinitrobenzene.

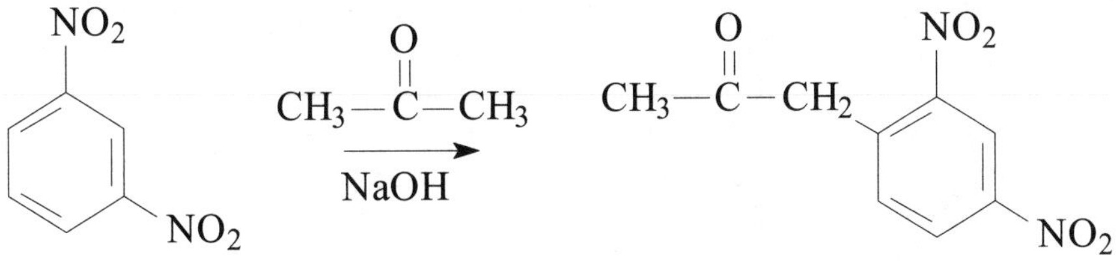

Figure 3.25.4 Formation of Janousky product.

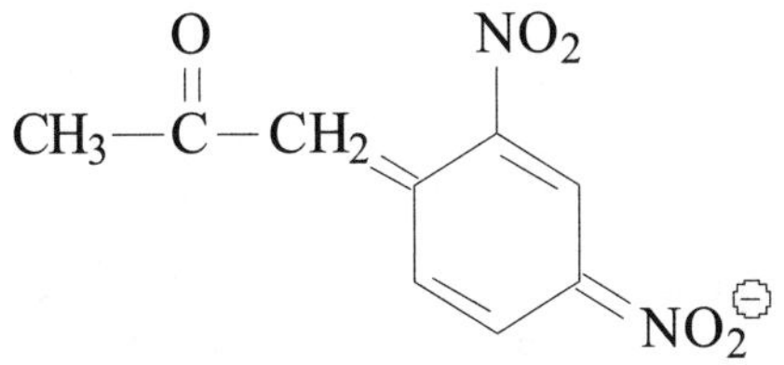

Figure 3.25.5 Zimmermann product.

In the second step, the acidic environment is made alkaline by adding 5 ml *strong sodium hydroxide solution R*. The sodium hydroxide should be added slowly and the reaction tube lowered into iced water due to a strong evolvement of heat. Then 5 ml of acetone is added, which upon mixing separates into two layers of which the upper is colored deep violet.

Acetone reacts with m-dinitrobenzole to form the violet Janovsky product (Figure 3.25.4). Water and acetone, which are normally miscible in all proportions, separates because of the salting-out effect of the high concentrations of ions in the water. The Janovsky product is in the upper acetone phase.

The formation of a colored product of methyl ketones by the reaction of m-dimethylketone has been widely used for colorimetric determinations of a large number of oxy compounds, and has for this reason been extensively studied. Compounds having a carbon with at least 1 H in ß position that will under alkaline conditions partly be present as the enolat ion. This enolat ion reacts with the m-dinitrobenzene by nucleofilic substitution.

In circumstances where the m-dinitrobenzene is in excess, it will oxidize the Janovsky product to the Zimmermann product (Figure 3.25.5). There will be an alkaline environment in equilibrium between the Zimmermann form and the Janovsky form. If the ketone is in excess, as in the present test, the Janovsky product will be the predominant form.[3]

A large number of compounds qualify, so the method has been used for determination of a large number of substances: methylethylketone, acetic acid, and several other acids and ketones.[4]

The method is quite specific because the two chemical reactions are combined. The second step is not selective as described above since only acetone is present and a contamination of the test solution with an organic compound of the discussed structure is unlikely. The reaction was been studied extensively to reveal which ions will cause interference. F^-, SO_4^{2-}, $C_2O_4^{2-}$, $C_4H_4O_6^{2-}$, SCN^-, SO_3^{2-}, CO_3^{2-}, $[Fe(CN)_6]^{\Omega-}$, NO_2^-, Ag^+, Hg^+, Bi^{3+}, Pb^{2+}, Cu^{2+}, Cd^{2+}, As^{3+}, Sb^{3+}, Sn^{2+}, Co^{2+}, Ni^{2+}, Fe^{2+}, Mn^{2+}, Al^{3+}, Cr^{3+}, Zn^{2+}, NH_4^+, as well as the alkali and earth-alkaline cations do not cause any problems.[5] A potential problem could be if the nitrobenzene reagent used is contaminated with dinitrobenzene.

A drawback of the method is the toxicity of the nitrobenzene and it is worth mentioning that the work of Werner has revealed that substitution to the less toxic 2-nitroethylbenzene and 4-nitroethylbenzene is possible.

References

1. Pesez, M.M., Sur une nouvelle réaction spécifique de l'acide nitrique et de ses sels, *J. Pharm. Chim.*, 29, 460, 1939.
2. Pesez, M.M., Sur un nouveau microdosage colorimétrique des nitrates, *J. Pharm. Chim.*, 30, 112, 1939.
3. Kakáč, B. and Vejdělek, Z.J., *Handbuch der photometrishe analyse organisher verbindungen*, Band 1, Verlag Chemie, Weinheim, 1974, p. 273.
4. Lange, B. and Vejdělek, Z.J., *Photometrische analyse*, 1st ed., Verlag Chemie, Wienheim, 1980, 487.
5. Werner, W., Der nitrat-nachweis von Pesez im DAD 9/Ph. Eur. 2, *Arch. Phar. (Weinheim)*. 323, 955, 1990.

3.26   Phosphates (Orthophosphates) (European Pharmacopoeia 2.3.1)

The test identifies the presence of phosphate ions, PO_4^{3+} in the test solution analyzed, originating either from a parent dissociated phosphate salt or from a chemically bound phosphate released by hydrolysis. In the heading it is specified in parentheses that orthophosphates is only used to distinguish them from other phosphates. Reference to phosphate is made in little less then 30 monographs, the majority describing phosphate salts. In about one-third of the monographs only test (b) is dictated and in the remaining test (a) or both together are dictated equally.

Phosphorus is capable of making many different oxoacids of which some of most important are: orthophosphoric acid, H_3PO_4. diphosphoric acid, $H_4P_2O_7$, which is a dimeric example of the series of orthophosphoric acid anhydrates called pyrophosphoric acid. Metaphosphoric acid is a polymeric form of phosphorous acid, HPO_3. Since these substances, and the other oxoacids, are unstable and fairly easily convert into orthophosphoric acid is by far the most widely encountered, and for the same reason it is normally referred to as just phosphoric acid without the prefix ortho.

Orthophosphates can be regarded as a series of corresponding salts of phosphoric acid.

$$Na^+ + H_3PO_4 \rightarrow Na^+ + NaH_2PO_4 + H^+ \rightarrow$$
$$Na^+ + Na_2HPO_4 + 2H^+ \rightarrow Na^+ + Na_3PO_4 + 3H^+$$

With the exception of lithium and ammonium all alkali metals and the primary salts of the alkali earth metals form water-soluble salts. All other salts are water insoluble.

This monograph contains two tests. Test (a) is a silver salt precipitation of limited selectivity, and test (b) is a far more specific test that is based on a precipitation with molybdate and vanadate.

Test (a)

In the first step of the analysis, 5 ml of the test solution is, following neutralization, added to 5 ml of *silver nitrate solution R1*. This gives a yellow precipitate of silver phosphate.

$$Ag^+ + HPO_4^{2-} \rightarrow Ag_3PO_4\downarrow + H$$

This differentiates orthophosphate from pyrophosphoric acid and metaphoric acid that forms white precipitates. Silicate, bromide, iodide, and iodate all give yellow or pale yellow precipitates under similar conditions.

In the second step, the precipitate is boiled and must not change appearance. This differentiates it from phosphorous acid and hypophosphorous acid, which would reduce Ag^+ to metallic silver, turning the precipitate black. In the last step of analysis *ammonia R* is added upon which the precipitate dissolves as the silver ion forms a complex with ammonia.

$$Ag_3PO_4\downarrow + 6NH_3 \rightarrow 3[Ag(NH_3)_2] + PO_4^{3-}$$

This rules out false positive reactions caused by the presence of bromide and iodide since they are insoluble to the extent that complexation will not take place. Arsenic(III) gives a reaction similar to phosphate but a content of this cation would be revealed in test (b). The high number of insoluble silver salt lowers the ruggedness of the method, since the presence of even low concentrations of organic and inorganic cations gives precipitates of various colors.[1]

Test (b)

This reaction, discovered by Misson in 1908, has found widespread use in quantitative determinations phosphor and phosphorus.

The test is simply to mix 1 ml of test solution with 2 ml of *molybdovanadic reagent R* upon which a yellow color is formed. The nature of the substance formed has not been fully elucidated and it has been referred to as an acid, a complex, and in some cases a salt. One reference gives the formula below.

$$(NH_4)_3PO_4 \cdot NH_4VO_3 \cdot 16MoO_3$$

It has been stated to be the most specific spectroscopic method available for phosphate analysis and was therefore widely used and dealt with in published literature. Since details in the test conditions described vary somewhat in-between the methods published, it is evident that the procedure has some inherent ruggedness. This applies especially when the methodology is used for identification purposes only, not measuring the absorbance at a specific wavelength and not comparing with a standard. In literature dealing with the method used for assays, the reagent and analyt concentrations vary in intervals stated below.

Reagent	Molar concentration
Acid	0.04–1.0
Vanadate	0.00008–0.0022
Molybdate	0.0016–0.057

The acids used in the methods are nitric, sulfuric acid, hydrochloric acid, and perchloric acid. If the acid concentration is very low, other precipitation may predominate, and if the concentration is too high, the color will be faint or ultimately obsolete. A maximum is reported to be at approximately pH 1.5. Even though the vanadate/molybdate concentration may vary a lot, vanadate should always be in excess to the analyt. In older publications vanadate and molybdate are added separately and in this case they should be added in the order acid, vanadate, and molybdate. Changing this apparently can lead to other precipitations or complexes.

A very thorough investigation by Kitson and Mellon reveals that hardly any ions cause interference and only chromate and dichromate ions were found to interfere in concentration equivalent to the phosphate concentration.[2] Reducing agents will reduce the complex to molybdenum blue, a feature actually used in some determinations. Organic ions such as citrate, tartrate, and oxalate will form a complex with molybdate. It should be kept in mind though that these compounds are reported as potential interferences in quantitative determination. Whether they could actually eliminate a false positive reaction in an identification is questionable. When used for quantitative determinations the color is reported to be stable for 3 h, while others report a stability of 24 h.[3–5]

A thorough review of the method, used in spectroscopic assays has been made by Rieman and Beukenkamp and includes more that 40 references.[6]

References

1. Reimers, F., *The basic principles for pharmacopoeial tests*, Heinemann Medical Books, London, 1956, p. 22.
2. Kitson, R.E. and Mellon, M.G., Colorimetric determination of phosphorus as molybdivanadophosphoric acid, *Industrial and engineering chemistry. Analytical edition*, 16, 379, 1944.

3. Kakáč, B. and Vejdělek, Z.J., *Handbuch der photometrishe analyse organisher verbindungen*, Band 1, Verlag Chemie, Weinheim, 1974, p. 688.
4. Lange, B. and Vejdělek, Z.J., *Photometrische analyse*, 1st ed., Verlag Chemie, Wienheim, 1980, p. 373.
5. Kakáč, B. and Vejdělek, Z.J., *Handbuch der photometrishe analyse organisher verbindungen*, 2. Ergänzungsband, Verlag Chemie, Weinheim, 1983, p. 308.
6. Kolthoff, I.M. and Elving, P.J., *Treatise on analytical chemistry*, Part II, Vol. 5, Interscience Publishers, New York, 1961, p. 317.

3.27 Potassium (European Pharmacopoeia 2.3.1)

The test identifies the substance to be examined as a salt of potassium, K^+. As with other alkali metals, potassium is exclusively found as the monovalent cation. Since almost all salts of potassium are water soluble and it is a very inert ion, as a rule these do not contribute to the characteristics of its salt. For this reason its function in the compounds enrolled in the pharmacopoeia is to be an indifferent counter-ion, both for simple inorganic anions but also for anionic active pharmaceutical ingredients. Reference to potassium is made in about twenty-five monographs. In about half of them test (b) is dictated in about one-quarter and in the last quarter both tests have to be performed. The anions in the organic salts are, with the exception of Acesulfame, carboxylic acids. Due to the high water solubility of potassium, sample preparation is in most cases just a simple dilution in water.

Of the few water in-soluble salts potassium can produce, precipitation with tartrate and cobolt nitrate respectively is used for the identification.

Test (a)

A 2 ml test solution, containing 0.1 g of the substance to be examined in water, is added to 1 ml of *sodium carbonate solution R*, and heat is applied. No precipitation must occur. This step reveals the presence of cations capable of forming insoluble carbonates.

In the next step, 0.05 ml of *sodium sulfide solution R* is added to the hot test solution, upon which no precipitation must occur. This reveals the presence of cations cable of forming insoluble sulfide salt even at neutral pH.

The test solution is finally cooled and ice is added to 2 ml of a 150 g/l solution of *tartaric acid R*. When allowed to stand a white crystalline precipitate of potassium hydrogen tartrate is formed (Figure 3.27.1).

The precipitation might have a slow onset but can be accelerated by scratching the inside of the test tube with a glass rod. It is soluble in both

$$K^+ \ + \ HOOC-\overset{OH}{\underset{|}{C}}-\overset{OH}{\underset{|}{C}}-COOH \ \longrightarrow \ H^+ \ + \ KOOC-\overset{OH}{\underset{|}{C}}-\overset{OH}{\underset{|}{C}}-COOH$$

Figure 3.27.1 Formation of potassium hydrogen tartrate.

strong acid and base. Tartaric acid will form insoluble salts with a relatively high number of cations, so the selectivity is given by the two preliminary precipitations. Adding sodium carbonate reveals the presence of calcium and other cations forming insoluble oxides and hydroxides. Adding sodium sulfide reveals the presence of lead and other heavy metals, including thallium which otherwise behaves much like potassium. The only cation that with certainty will give a false positive reaction is ammonium since it does not precipitate with carbonate or sulfide and affords a tartrate salt that is very similar to the potassium salt.[1]

Test (b)

A test solution is prepared by dissolving about 40 mg of the substance to be examined in 1 ml of *water R*, or by using the 1 ml of the prescribed solution. Then, 1 ml of *dilute acetic acid R* and 1 ml of a freshly prepared 100 g/l solution of *sodium cobaltnitrite R* is added. A yellow or orange-yellow precipitate is formed immediately. The exact composition of the complex responsible for the precipitate varies with test solution sodium and potassium concentrations, of test solution pH, and temperature.

If limited amounts of sodium are present, the pure potassium complex will prevail.

$$3K^+ \, [Co(NO_2)_6]^{3-} \rightarrow K_3[Co(NO_2)_6]\downarrow$$

If sodium is more abundant, a mixed complex is formed with the composition $K_2Na[Co(NO_2)_6]$. The two complexes are alike in appearance. This reaction was used in the classical inorganic analysis to isolate potassium from the other alkali metals. The precipitate was isolated and washed and a determination was then performed colorimetrically on the cobalt ion. In many cases the silver, zinc, or lead salt of hexanitritocobaltate was used as a reagent.

There are some differences in literature about which ions are claimed to interfere with the determination. Some state that ammonium, thallium, and rubidium give precipitates that would cause a false positive identification, and the reaction is actually used in 2.6. Ammonium salts. One reference adds caesium, barium, zirconium, lead, and mercury, but without stating the nature of interference.[1] Some references describe an analogue method where $AgNO_3$ is added that give a precipitation of $K_2Ag[Co(NO_2)_6]$, but this can only be used when the test solution is halide free. In this method, lithium is said to give a precipitate as well, but whether this is the case when precipitation is done without silver as well is not clear.[2] Presumably, this is not the case of one reference state that the lithium salt of cobalt nitrite can be used as a reagent.[3] If the test solution by mistake or contamination is alkaline the reagent will be destroyed by precipitation of black cobalt(III) hydroxide. The test is also mentioned to be sensitive toward reducing substances. The sodium cobaltnitrite reagent has to be prepared fresh since its sensitivity is reduced on standing.[4,5]

Only the lack of specificity toward ammonium is a real concern, primarily since test (a) also gives a positive reaction with this ion, and second because thallium and rubidium hardly are realistic contaminants.

References

1. Kolthoff, I.M. and Elving, P.J., *Treatise on Analytical Chemistry*, Part II, Vol. 1, Interscience Publishers, New York, 1959, pp. 361, 370.
2. Feigl, F., *Qualitative Analysis by Spot Tests*, 3rd ed., Elsevier Publishing Co., New York, 1947, p. 173.
3. Kodama, K., *Methods of Quantitative Inorganic Analysis*, Interscience Publishers, New York, 1963, p. 409.
4. Lange, B. and Vejdělek, Z.J., *Photometrische analyse*, 1st ed., Verlag Chemie, Wienheim, 1980, p. 133.
5. Allport, N.L., *Colorimetric Analysis*. 1st ed., Chapman & Hall, London, 1945, p. 98.

3.28   Salicylates (European Pharmacopoeia 2.3.1)

The test identifies the substance to be examined as a salicylate or a derivative that yields a salicylate upon hydrolysis. A salicylate is the corresponding base of salicylic acid (Figure 3.28.1) or a derivative hereof.

Salicylic acid and a number of related substances are used in therapy as nonsteroid antiinflammatory drugs. Besides the acetyl ester (Figure 3.28.2) aspirin, 5-amino salicylic acid (Figure 3.28.3) could be mentioned. Salicylic acid is also used for its acidic properties in the treatment of warts because it is a relatively strong organic acid with a pKa of 3. At the present reference to salicylates is made in six monographs including salicylic acid itself and its sodium salt. Being a carboxylic acid, salicylic acid has a water solubility that depends strongly on solvent pH. This, and its well-defined melting point, forms the basis of test (b). In test (a) a strongly colored complex is formed with iron(III).

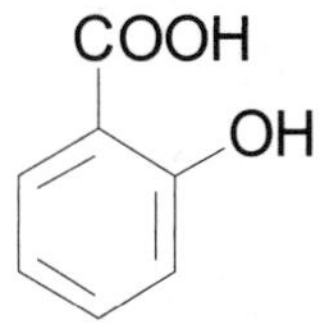

Figure 3.28.1 Salicylic acid.

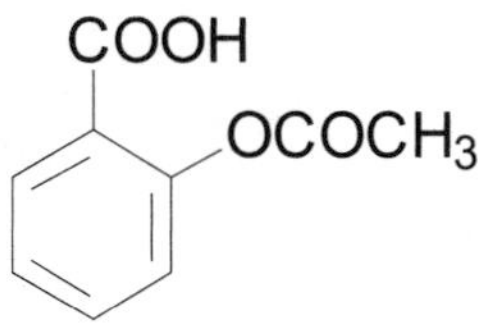

Figure 3.28.2 Acetyl salicylic acid.

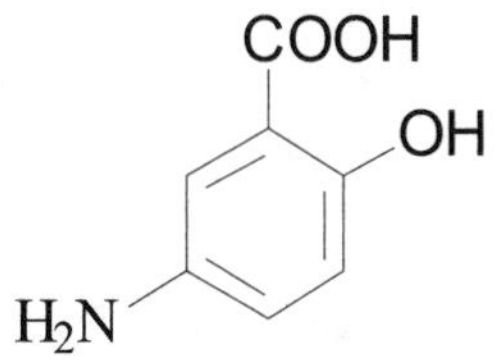

Figure 3.28.3 5-amino salicylic acid.

ml of the prescribed solution is added to 0.5 ml of *ferric chloride solution R1*. This gives a violet color, which persists after the addition of 0.1 ml of *acetic acid R*.

Phenolic compounds in general give a blue, violet, or red complex with iron(III) ions, and being a phenol derivative salicylate behaves in the same manner. When phenols are substituted with CH_3, OH, NH_2, or a halogen at the meta-position, they give a relative stable blue to violet complex, but when substituted at the ortho or para position they do not. When substituted with CHO, COOH, COOR, NO_2, or SO_3H, they give a purple to red color. In the case of carboxylic substituents, meta and para positions give an unstable complex whereas the ortho position gives a stable complex. The color and the stability of the complexes are also dependent upon the pH of the solution and the presence of solvents such as the lower alcohols.

Esters of salicylic acid have to be hydrolyzed before testing to make the carboxylic acid and alcohol group available for complexation.

Compounds other than phenols are also capable of giving color reactions with iron(III). Of special interest are the ones forming red or violet complexes: formate and acetate, benzoate, acetylacetone, and antipyrine. Even some steroids can give red complexes with iron(III) through enolization of a carbonyl group.[1-3] Selectivity in the test method is given by the color of the complex and its stability even after adding acetic acid.

Test (b)

Dissolve 0.5 g of the substance to be examined in 10 ml of water R or use 10 ml of the prescribed solution. Add 0.5 ml of *hydrochloric acid R*. The precipitate obtained, after recrystallization from hot *water R* and drying in vacuo, has a melting point (2.14) of 156°C to 161°C.

The salicylic acid, hydrolyzed if necessary, precipitates when pH is lowered so that the molecule is brought at the acid form. One reference states that recrystallization is sometimes necessary to obtain the good result.[4]

References

1. Pesez, M. and Bartos, J., Colometric and Fluorimetric Analysis of Organic Compounds and Drugs, Marcel Dekker, New York, 1974, p. 78.
2. Vejdělek, Z.J. and Kakáč, B., Farbreaktionen in der spectrophotometrischen analyse organischer verbindungen, Band II, VEB Gustav Fischer Verlag, Jena, 1973, p. 356.
3. Lange, B. and Vejdělek, Z.J., Photometrische analyse, 1st ed., Verlag Chemie, Wienheim, 1980, p. 542.
4. Hartke, K., 2.3.1. Identitätsreaktionen (10. Lfg.1999), in Arzneibuch-Kommentar zum europäische arzneibuch, Band I Allgemeiner teil, Hartke, K. et al., Wissenschaftliche Verlagsgellschaft, Stuttgart/Govi-Verlag — Pharmazeutischer Verlag, Escbor.

3.29 Silicates (European Pharmacopoeia 2.3.1)

The test identifies that the substance to be examined contains silicon in a form that can be converted into a silicate. A silicate is defined as "Any complex anion, or the salt containing it, that possesses a central silicon atom; in particular the ion SiO_4^{4-} and its derivatives. Reference to silicates is made in 13 monographs and they can roughly be divided into silicium salts or minerals and silicones. Silicon itself is a nonmetallic element and therefore does not form simple ions. But it has a strong tendency of forming both neutral and anionic compounds with oxygen, for example SiO_2, SiO_4^{4-}, and $Si_2O_5^{2-}$. These anions occur in a very large number of minerals of which talc, $Mg_3(Si_4O_{10})(OH)_2$ that is used as a glidant, is an example. Silicones are polymers or oils containing the basic unit $O-SiR_2-$ where R is an alkyl substituent.

Since the majority of silicon salts are insoluble in water the anion first has to be liberated from the substance to be examined by incineration, often with an alkali carbonate. Silicone is purely covalent and has to be converted into a silicate by incineration. The ability of silicate to form a volatile complex with fluoride and the water insolubility of silicon dioxide is used in the identification.

First, the substance (maybe pretreated) to be examined is mixed with about 10 mg *sodium fluoride R* and a few drops of *sulfuric acid R* to give a thin slurry. This leads to the formation of hydrofluorosilicic acid as shown below using silicon dioxide as an example.

$$SiO_2 + 6HF \rightarrow H_2[SiF_6] + 2H_2O$$

The analysis must be performed in a lead or platinum crucible, since glass and silica crucibles are silica based. In the second step, the crucible is covered with a thin transparent plate of plastic under which a drop of water is suspended and heat is applied. As a response to this, gaseous (boiling point 90°C) silicium fluoride is liberated.

$$H_2[SiF_6] + heat \rightarrow SiF_4\uparrow$$

Fluoride should be present in a less then equivalent amount since the interfering H_2F_2 could be formed if present in excess. If heating is done in a lead crucible, care should be exercised since lead melts at 327°C.

When the gas comes in contact with the water drop, the following reaction occurs.

$$3SiF4\uparrow + 2H_2O \rightarrow SiO_2\downarrow + 2[SiF_6]^{2-} + 4H^+$$

The precipitation of silicon dioxide gives a white ring around the drop of water.[1,2] Some references ascribe the precipitate as silicic acid, which may be presented by the formula, $SiO_2(H_2O)_{n-2}$.[3]

References

1. Vogel, A.I. and Svehla, G., *Vogel's Qualitative Inorganic Analysis*, 6th ed., Longman Scientific & Technical, Essex, 1987, p. 196.
2. Charlot, G., *Qualitative Inorganic Analysis. A New Physio-Chemical Approach*, Methuen & Co., London, 1954, p. 317.
3. Hartke, K., 2.3.1. Identitätsreaktionen (10. Lfg.1999), in *Arzneibuch-Kommentar zum europäische arzneibuch*, Band I *Allgemeiner teil*, Hartke, K. et al., Wissenschaftliche Verlagsgellschaft, Stuttgart/Govi-Verlag — Pharmazeutischer Verlag, Escbor.

3.30   Silver (European Pharmacopoeia 2.3.1)

The test identifies the substance to be examined as a salt of silver (Ag^+). At the present silver is referenced in only one monograph, silver nitrate. This salt is sold in sticks, called Lapis or Lapis lunaris, and is used for the treatment of warts. Silver can exist both as silver(I) and silver(II), but since the latter is less stable, silver(I) dominates. It is a noble metal meaning that it is most stable in oxidation state 0, its metallic form. So, contrary to most metals, it does not have at tendency to be oxidized to its ionic form and thereby corrode. Silver nitrate and fluoride are soluble, while silver nitrate, acetate, and sulfate have limited solubility. All other salts are insoluble but are, however, capable of forming many soluble complexes. The insolubility of the chloride salt and its ability to form a soluble nitrate complex are useful in identification.

In the first step of the analysis, the 10 ml test solution containing about 10 mg of the substance to be examined is added to 0.3 ml of *hydrochloric acid R1*. This gives a white curdled precipitate of silver chloride.

$$Ag^+ + Cl^- \rightarrow AgCl\downarrow$$

Only a limited number of metals form insoluble chlorides, see periodic table below, and of these only the precipitates of silver, mercury, lead, and thallium are white.

H																	He
Li	Be											B	C	N	O	F	Ne
Na	Mg											Al	Si	P	S	Cl	Ar
K	Ca	Sc	Ti	V	Cr	Mn	Fe	Co	Ni	Cu	Zn	Ga	Ge	As	Se	Br	Kr
Rb	Sr	Y	Zr	Nb	Mo	Tc	Ru	Rh	Pd	Ag	Cd	In	Sn	Sb	Te	I	Xe
Cs	Ba	La	Hf	Ta	W	Re	Os	Ir	Pt	Au	Hg	Tl	Pb	Bi	Po	At	Rn
Fr	Ra	Ac	Th	Pa	U												

In the next step, 3 ml of *dilute ammonia R1* is added and the precipitate dissolves. This is due to Ag^+ ions' ability to form the diammonium-argentate complex.

$$AgCl\downarrow + 2NH_3 \rightarrow [Ag(NH_3)_2]^+ + Cl^-$$

Being given the same treatment, $PbCl_2$ would not redissolve but possibly change into another insoluble salt, lead hydroxide.

$$PbCl_2\downarrow + 2NH_3 + 2H_2O \rightarrow Pb(OH)_2\downarrow + 2NH_4^+ + 2Cl^-$$

Hg_2Cl_2 would disproportionate into Hg and Hg^{2+} giving a shiny black appearance due to the presence of metallic mercury.

$$Hg_2Cl_2\downarrow + 2NH_3 \rightarrow Hg\downarrow + Hg(NH_2)Cl\downarrow + NH_4^+ + Cl^-$$

Therefore, of the ions forming white insoluble chlorides, only thallium cannot be excluded as a cause of a false positive reaction. The risk of finding thallium in a pharmaceutical sample must be considered very small and the same could be said about another possible cause of a false positive reaction, the tungstate ion. By the addition of dilute hydrochloric acid it will form, not an insoluble chloride salt, but a white precipitate of hydrated tungstic acid.

$$WO4^{2+} + H^+ \rightarrow H_2WO_4$$

In general, fairly good selectivity is obtained, especially compared to the selectivity obtained when the same precipitation is used for identification of the different halides, as for example in 3.16. Chlorides, 3.13. Bromides, and 3.19. Iodides.

The presence of an even mildly reducing substance would eliminate the Ag^+ reagent by turning it into metallic silver. Presence of any of the anions that precipitate with the silver obviously interferes. If, by mistake, the test is performed using concentrated hydrochloric acid, no precipitation will occur due to the formation of a dichloroargentate complex.

$$AgCl\downarrow + Cl^- \rightarrow [AgCl_2]^-$$

If the precipitate is exposed to light it will turn grayish or black as Ag^+ is reduced to the metal.[1,2]

$$2AgCl\downarrow + \lambda \rightarrow 2Ag + Cl_2$$

References

1. Vogel, A.I. and Svehla, G., *Vogel's Qualitative Inorganic Analysis*, 6th ed., Longman Scientific & Technical, Essex, 1987, several chapters.
2. Holness, H., *Advanced Inorganic Qualitative Analysis by Semi-Micro Methods*, 1st ed., Sir Isaac Pitman & Sons, London, 1957, p. 120.

3.31 Sodium (European Pharmacopoeia 2.3.1)

The test identifies the substance to be examined as a salt of sodium, Na^+. Reference to sodium is made in more than 100 monographs. Out of these test (a) is dictated in about 70 monographs, only two monographs dictate test (b), and 15 monographs dictate both. Being an alkali metal, sodium is exclusively found as the monovalent ion. Very few sodium salts are insoluble and sample preparation is in most cases done simply by dissolving the substance to be examined in water. In the majority of monographs sodium plays the role of an indifferent counterion, regardless of whether the anion is a large organic molecule or a simple inorganic ion.

The classic reagent for qualitative and quantitative determination of sodium is uranyl acetate, which in the presence of sodium and a divalent metal cation gives a precipitate of $NaM(UO_2)_3(CH_3COO)_9$, where M could be Mg^{2+}, Zn^{2+}, Co^{2+}, Ni^{2+} or others. However, since uranium is very poisonous and radioactive, tests based on these reagents have been replaced with tests based on sodium's reactivity toward pyroantimonate and methoxyphenylacetic.

Test (a)

To prepare the test solution, 0.1 g of the substance to be examined is dissolved in 2 ml of *water R*, or 2 ml of the prescribed solution is used. 2 ml of a 150 g/l of *potassium carbonate R* is added and the test solution is boiled. No precipitate is formed.

In the next step of the analysis, 4 ml of *potassium pyroantimonate solution R* is added and the test solution is boiled. The test solution is allowed to cool in iced water and if necessary the inside of the test tube is rubbed with a glass rod. A dense white precipitate of sodium antimonate is formed.

$$Na^+ + [Sb(OH)_6]^- \rightarrow Na[Sb(OH)_6]\downarrow$$

This salt has a tendency of forming rather stable super-saturated solutions, and this is the reason the test solution has to be cooled down and precipitation initiated by rubbing with a glass rod. For the same reason sodium hydroxide is added to the *potassium pyroantimonate solution R* in the reagent preparation so that saturation is almost reached. Both precautions are necessary to achieve a sufficient sensitivity.[1]

The purpose of adding potassium carbonate in the first step of analysis is twofold. First it raises the test solution pH thereby preventing antimonate precipitates as antimonic acid, a reaction that could occur in neutral and weakly acidic solutions.[2]

$$Sb^{5+} + 4H_2O \rightarrow H_3SbO_4\rightarrow + 5H^+$$

The second purpose is to reveal the presence of cations that would otherwise give a false positive reaction by precipitating them as carbonates. Calcium is an example of such a cation. This first step has been the object of some debate since it has been observed that it is obviously possible to perform the pyroantimonate precipitation on a substance even though addition of carbonate has been omitted.[3] The success of the test will, however, depend on the absence of, for example, calcium. Sodium's close neighbor in the alkali metal group, lithium, is claimed to give a false positive reaction.[4]

Test (b)

The test is based on a methodology first published by Reeve and Chistoffel in 1957. The first two steps of analysis as described in the European *Pharmacopoeia* are performed exactly as proposed in the original article, which gives a very detailed discussion about the strengths and weaknesses of the procedure.[5]

The test solution is prepared by dissolving a quantity of the substance to be examined equivalent to about 2 mg of sodium (Na$^+$) in 0.5 ml of *water R*, or by using the prescribed solution. In the first step of analysis 1.5 ml of *methoxyphenylacetic reagent R* is added to the test solution, and it is cooled in ice water for 30 min. A voluminous, white, crystalline precipitate is formed. The precipitate is the sodium salt of methoxyphenylacetic acid (Figure 3.31.1).

The sodium to methoxyphenylacetic acid ratio in the salt is apparently 1:2. The precipitation has an optimum in the pH interval 3 to 4 and this is achieved by the reagent. At pH values below this interval no precipitation will take place. The lower limit of the determination is 0.3 mg sodium per ml but in such cases the precipitation might take 1 h to initiate.

The test tube is moved to a 20°C water bath and stirred for 5 min, and this procedure must not dissolve the precipitate. Many of the other alkali metals, in addition to ammonia, will also precipitate as a methoxyphenylacetic acid salt if present in a high concentration. But contrary to the sodium salt they will redissolve when transferred to this higher temperature. It should be noted, however, that "high concentrations" in this context equal as much as 8 M in the case of magnesium and 2.3 M in the case of potassium. Cesium will interfere in the test if present in the test solution in a concentration of 10 mg/ml. Lithium is tolerated in 30 mg/ml.

In the third step of the analysis, 1 ml of dilute ammonia is added upon which the precipitate dissolves. And in the final step of analysis 1 ml of *ammonium carbonate solution R* is added. This must not give a precipitate. This alkaline media will dissolve the sodium methoxyphenylacetic acid salt, but reveal if the precipitate was of one of the earth alkali metals calcium, strontium, or barium, since these ions form insoluble hydroxides and carbonates at high pH. Also salts of heavy metals are revealed in this way.[6–8]

Overall, it can be concluded that no other cations are capable of giving a false positive reaction, but quite a large number of ions could cause interference in the test by forming a precipitate in one of the precipitations aiming at enhancing the selectivity of the test.

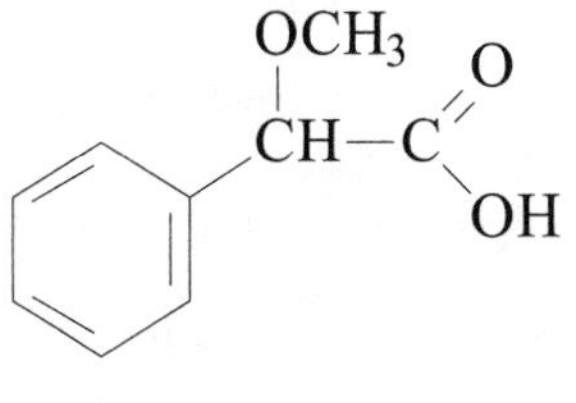

Figure 3.31.1 Methoxyphenylacetic acid.

References

1. Hartke, K., 2.3.1. Identitätsreaktionen (10. Lfg. 1999), in *Arzneibuch-Kommentar zum europäische arzneibuch*, Band I *Allgemeiner teil*, Hartke, K. et al., Wissenschaftliche Verlagsgellschaft, Stuttgart/Govi-Verlag — Pharmazeutischer Verlag, Escbor.
2. Vogel, A.I. and Svehla, G., *Vogel's Qualitative Inorganic Analysis*, 6th ed., Longman Scientific & Technical, Essex, 1987, p. 94.
3. Council of Europe, *Pharmeuropa*, September 1989, Council of Europe, Strasbourg, 1989, p. 234.
4. Reimers, F., *The Basic Principles for Pharmacopoeial Tests*, Heinemann Medical Books, London, 1956, p. 23.
5. Reeve, W., Chistoffel, I., Alpha-methoxyphenylacetic acid as reagent for detection of sodium ion, *Analytical Chemistry*, 29, 102, 1957.
6. Werner, W., DAB 9: -Methoxyphenylessigsure als reagenz auf natrium, *Pharmaceutiche Zeitung*, 134, 55, 1989.
7. Werner, W., Qualitative anorganishe analyse, 2. Auflage, Gustav Fischer Verlag, Stuttgart, 1997, p. 112.
8. Schorn, P.J. (Ed.), *Pharmeuropa*, December 1990, Council of Europe, Strasbourg, 1990, p. 218.

3.32   *Sulfates (European Pharmacopoeia 2.3.1)*

The test identifies that the substance to be examined is a salt of sulfate, SO_4^{2-}, or a substance that yields sulfate ions upon the chemical manipulations of the sample preparation. Reference to sulfates is made in about 60 monographs and in about two-thirds of these test (a) or both tests are used without chemical cleavage or modification, except from measures to enhance solubility. Test (b) is performed on the suspension obtained in test (a) and its purpose is to exclude any false positive reactions in test (a). Besides the multiple simple inorganic sulfate salts of the pharmacopoeia, many active pharmaceutical ingredients that are organic nitrogen bases are used as sulfate salt, owing to their relatively high water solubility.

Its oxides and oxo acids dominate the inorganic chemistry of sulfur. The most important of these are the oxides of sulfur(VI) and sulfur(IV), sulfur trioxide, and sulfur dioxide. They react with water to form sulfuric acid and sulfurous acid, respectively.

$$SO_3 + H_2O \rightarrow H_2SO_4$$

$$SO_2 + H_2O \rightarrow H_2SO_3$$

As also the oxides of phosphor, they form polymers willingly. Sulfuric acid is a strong acid, a dehydrating agent and an oxidizing agent. Most inorganic salts of the anion sulfate are water soluble. One exception is barium sulfate and the characteristic of this salt is used in the test. The identification test for sulfates is special since the purposes of the different steps of the analysis are explained in parentheses in the pharmacopoeial text.

Test (a)

A test solution is made by dissolving about 45 mg of the substance to be examined in 5 ml of *water R* or by using 5 ml of the prescribed solution. Add 1 ml of *dilute hydrochloric acid R* and 1 ml of *barium chloride solution R1*. A white precipitate of barium sulfate is formed.

$$Ba^{2+} + SO_4^{2-} \rightarrow BaSO_4\downarrow$$

This precipitate is colloidal and gives a fairly stable suspension. The size of the crystals and the appearance of the suspension is somewhat dependent on the exact way the sulfate test solution and the barium reagent are mixed. Sulfite and dithionit ($S_2O_4^{2-}$) will give a similar reaction and iodate, selenate, and wolframate will do likewise in certain concentrations, and in cases where it is relevant to exclude these, the suspension obtained is used as a test solution for test (b).

Test (b)

In the first step of the analysis, to the suspension obtained during reaction (a), 0.1 ml of *0.05 M iodine* is added. The suspension remains yellow (a distinction from sulfites and dithionites). If the white precipitate obtained in test (a) had been sulfite or dithionite they would reduce iodine to iodide, eliminating the yellow color.

In the second step, *stannous chloride solution R* is added drop-wise (distinction from iodates). This should decolorize the solution by reducing the iodine to iodide.

This will happen unless the white precipitate obtained in the first step of analysis was barium iodate. In this case iodate is reduced to iodate which shifts the above equilibrium to the left thereby preventing the iodine from being formed.

In the last step, the mixture is boiled. No colored precipitate is formed (distinctions of selenates and tungstates). Had the white precipitate been barium selenates or barium tungstates, they would have been reduced to elemental selenium and wolfram blue, respectively.[1,2]

References

1. Hartke, K., 2.3.1. Identitätsreaktionen (10. Lfg. 1999), in *Arzneibuch-Kommentar zum europäische arzneibuch*, Band I *Allgemeiner teil*, Hartke, K. et al., Wissenschaftliche Verlagsgellschaft, Stuttgart/Govi-Verlag — Pharmazeutischer Verlag, Escbor.
2. Reimers, F., *The Basic Principles for Pharmacopoeial Tests*, Heinemann Medical Books, London, 1956, p. 23.

3.33   Tartrates (European Pharmacopoeia 2.3.1)

The test identifies the substance to be examined to be a salt of tartaric acid. About 10 monographs references tartrates at the present. In the monograph for tartrate itself, both test (a) and test (b) are dictated, and in potassium hydrogen tartrate only test (a) is dictated. In all other cases, only test (b) is demanded. Monographs of a number of alkali metal tartrate salts are included in the pharmacopoeia, but they are used mainly as buffers in liquid pharmaceutical formulations. But besides this, tartrates have found some use as indifferent counter ions in salts of active pharmaceutical ingredients, predominantly organic nitrogen bases. Far from all monographs describing an organic tartrate ions make use of the general tartrates chapter identification, presumably because of problems with interference from the organic cation on the analytical procedure.

Tartrates are salts of the diprotic carboxylic acid (+)-tartaric acid (Figure 3.33.1) (or L-tartaric acid). This tartaric acid is the natural occurring form, but the (–)-tartaric acid and the racemate are commercially available as well. The two acid groups have pKa values of 2.93 and 4.43, respectively. The tartaric acid is a crystalline solid, with a high water solubility. Their ammonium and potassium salt has a limited solubility, a fact that in the pharmacopoeia is used for the identification of potassium in 3.27. Potassium. The salts of the other alkali metals are soluble, but salts of most other inorganic cations are sparingly soluble.[1]

Two different colorimetric tests are used in the pharmacopoeia for identifying tartrates, the Fenton test and the Pesez test.

Test (a)

In this test described by Fenton[1] in 1896, tartaric acid is oxidized to a substance which gives a color reaction with iron(III) ions.

A test solution is prepared by dissolving about 15 mg of the substance to be examined in 5 ml of *water R*, or 5 ml of the prescribed solution is used. 0.05 ml of a 10 g/l solution of *ferrous sulfate R* and 0.05 ml of *dilute hydrogen peroxide solution R* is added. A transient yellow color is produced, which disappears after some time.

Adding hydrogen peroxide to the test solution containing iron(II) ions oxidizes the tartaric acid to dihydroxfumeric acid. The reaction is a radical

$$
\begin{array}{c}
\text{COOH} \\
| \\
\text{HCOH} \\
| \\
\text{HOCH} \\
| \\
\text{COOH}
\end{array}
$$

Figure 3.33.1 Tartaric acid.

reaction where the hydrogen peroxide oxidizes iron(II) to iron(III) which produces a free radical.[2]

$$Fe^{2+} + H_2O_2 \rightarrow Fe^{3+} + OH^- + HO\cdot$$

This radical reacts with the tartaric acid as shown below to produce dihydroxyfumeric acid (Figure 3.33.2).

There has been some dispute however, whether some of the corresponding a cis isomers dihydroxymaleic acid (Figure 3.33.3), but this seems to have been ruled out.[3] The yellow color apparently comes from the colored complex of tartaric acid itself with iron before this is oxidized.[4]

Step two of the analysis is performed when the yellow color has disappeared. Here dilute *sodium hydroxide solution R* is added dropwise, and an intense blue color is produced caused by a complex between the dihydroxfumeric acid and the iron(III) ions. No structure of this complex has apparently been published. It is different from the general complexes normally formed between the aliphatic dihydroxy carboxylic acid with iron(III), since they are often yellow or violet and have an optimum in the pH region below neutrality. One could suspect that an enolization of the dihydroxfumeric acid is responsible for the blue colored complex,

It is stated that the reaction is not given by sugars, citric, succinic, malic, or oxalic acids. Racemic tartrates give a coloration of half the intensity of the l-tartaric acid. Calcium and phosphate apparently disturb the determination.[5] Since a high number of substances give colored complexes with iron(III) ions, selectivity of the test is given by the ability of the substance to be examined to behave exactly as expected in the sequence of procedures. It is not possible to make a list of compounds that could cause false positive reactions.

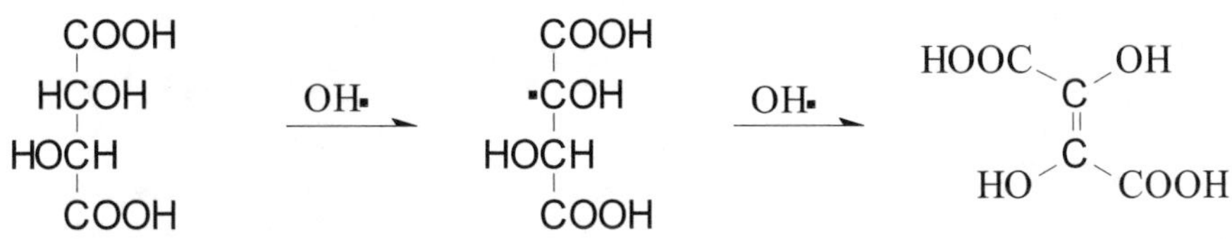

Figure 3.33.2 Radical reaction.

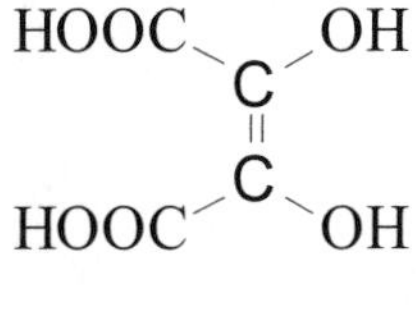

Figure 3.33.3 Cis dihydroxymaleic acid.

Test (b)

The test is based on a procedure published by Pesez in 1935,[6] in which an oxidative degradation product of tartaric acid is condensed with a reagent to make a colored substance.

In the first step of analysis, a test solution is prepared containing the equivalent of about 15 mg of tartaric acid per milliliter, or by following what is prescribed in the individual monograph. 0.1 ml of this solution is added to 0.1 ml of a 100 g/l solution of *potassium bromide R*, 0.1 ml of a 20 g/l solution of *resorcinol R*, and 3 ml of *sulfuric acid R*. The solution is heated on a water-bath for 5 to 10 min. A dark blue color develops.

The sulfuric acid oxidizes the tartaric first into glycol aldehyde and formic acid and finally the glycol aldehyde into glyoxylic acid (Figure 3.33.4).

The glyoxylic acid formed condenses with the resorcinol added, forming the lactone shown below (Figure 3.33.5). This is the compound responsible for the dark-blue color developed. The compound is insoluble in water and in acidic solutions.

In the second step, the test solution is allowed to cool and *water R* is poured into the solution. The color changes to red as the product of (Figure 3.33.5) is further oxidized and brome is added when the reaction shown below takes place (Figure 3.33.6).

In some reaction schemes for the identification of tartrate and for the general determination of aliphatic α,β-dihydroxycarboxylic acid, the same chemistry is applied but without the addition of brome. The result of this reaction is (Figure 3.33.7), which is stated to give a purple-red to orange-red color.[7,8]

The test will give a positive reaction with all substances, which, as tartaric acid, yields glyoxylic acid. This includes in principle all α,β-dihydroxy

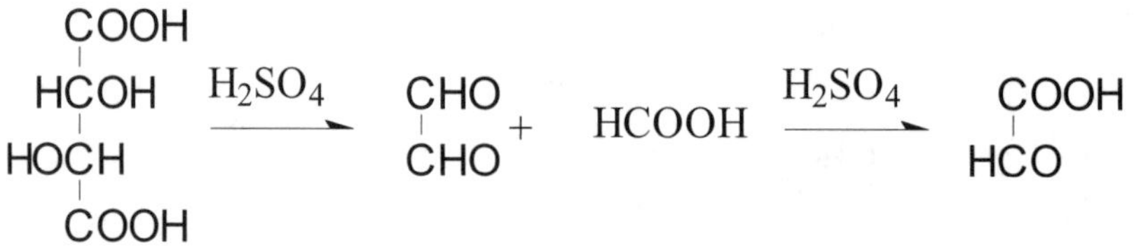

Figure 3.33.4 Glyoxylic acid formation.

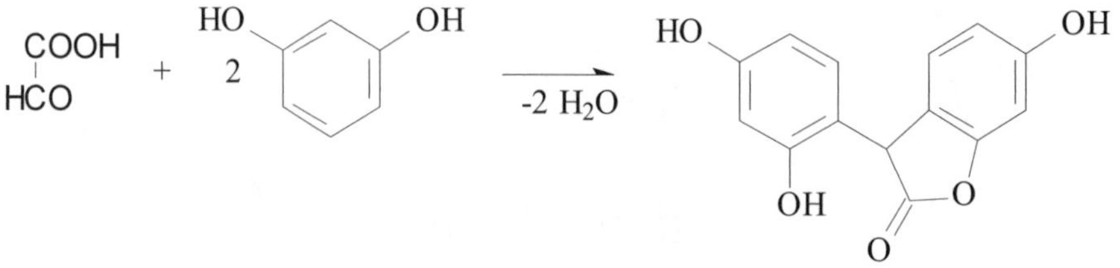

Figure 3.33.5 Condensation with resorcinol.

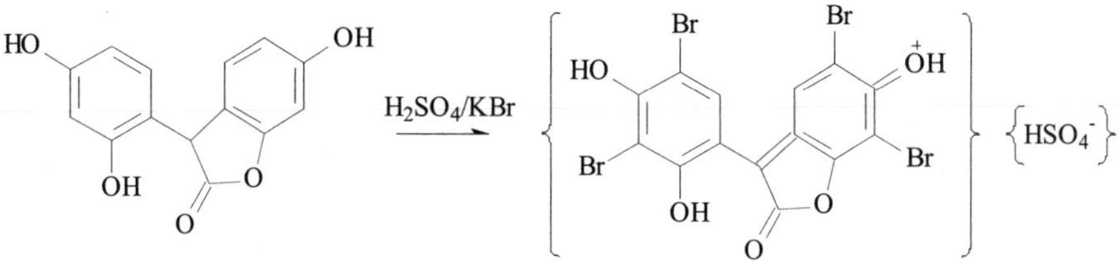

Figure 3.33.6 Addition of brome.

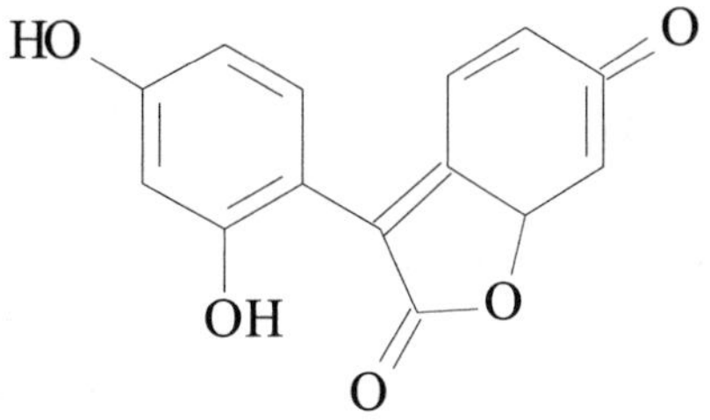

Figure 3.33.7 Reaction product.

carboxylic acids, although some without a doubt would require a stronger oxidant than sulfuric acid. This excludes the other lactates and citrates which in the *European Pharmacopoeia* is identified through different tests. No other substances than glyoxylic acid are capable of giving the condensation.

In references dealing with the above mentioned procedure, using periodate as an oxidant and not using bromate, it is stated that formaldehyde, acetic aldehyde, glycolic acid, oxalic acid, formic acid, amlone dialdehyde, and glyoxal do not interfere.

References

1. Fenton, H.J., The constitution of a new dibasic acid, resulting from the oxidation of tataric acid, *J. Chem. Soc.*, 69, 546, 1896.
2. Walling, C., Fentons reagent revisited, *Acc. Chem. Res.*, 8, 125, 1975.
3. Philippi, I. and Auterhoff, H., Zur kenntnis der tartratnachweise des europäischen arzneibuches, *Dtsch. Apoth. Ztg.*, 116, 205, 1976.
4. Pesez, M. and Bartos, J., *Colometric and Fluorimetric Analysis of Organic Compounds and Drugs*, Marcel Dekker, New York, 1974, pp. 61, 302.
5. Allport, N.L., *Colorimetric Analysis.* 1st ed., Chapman & Hall, London, 1945, p. 178.
6. Pesez, M., Sur une nouvelle réaction de l`acide tartrique, *J. Pharm. Chem.*, 21, 542, 1935.
7. Werner, W., *Qualitative anorganishe analyse*, 2. Auflage, Gustav Fischer Verlag, Stuttgart, 1997, p. 158.
8. Kakáč, B. and Vejdělek, Z.J., *Handbuch der photometrishe analyse organisher verbindungen*, Band 1, Verlag Chemie, Weinheim, 1974, p. 353.

3.34   Xanthines (European Pharmacopoeia 2.3.1)

The test identifies the substance to be examined as a member of the group of compounds called the xanthines, and at the present about 10 monograph reference xanthines. Xanthines are compounds that historically have been extracted from plant material containing the core element of xanthine (Figure 3.34.1). Biochemically they are derived from the nucleotide purine (Figure 3.34.2).

The different xanthines have various alkyl groups attached to the three nitrogen atoms not participating in the double bond. Examples are given below.

Compound	R_1	R_2	R_3
Caffeine	CH_3	CH_3	CH_3
Theophylline	CH_3	CH_3	H
Theobromine	H	CH_3	CH_3
Pentoxifylline	CH_3	$(CH_2)_4COCH_3$	CH_3

The xanthines are alkaloids, but they do not react to the classical alkaloid reagents that are discussed in 3.3 Alkaloids.

The test is based on a partly specific oxidation called the murexide reaction. In the first step of the analysis a few milligrams (or the prescribed quantity) of the substance to be examined are added to 0.1 ml of strong *hydrogen peroxide solution R* and 0.3 ml of dilute *hydrochloric acid R*. The mixture is heated to dryness on a water-bath until a yellowish-red residue is obtained.

In the second step of the analysis, 0.1 ml of dilute *ammonia R2* is added. The color of the residue changes to violet-red. The compound responsible for the coloration, murexoin, is an ammonium salt of the structure given below (Figure 3.34.3). It is named after the structurally related metal indicator murexide (Figure 3.34.4) and like this its color depends on the presence of cations in the solution.

A fairly large number of structurally-related compounds give a positive murexide reaction. They are useful as an identification reaction for uric acid (Figure 3.34.5) and other purine derivates. These substances all contain the same structural feature as the six-membered heterocyclic ring of xanthine.

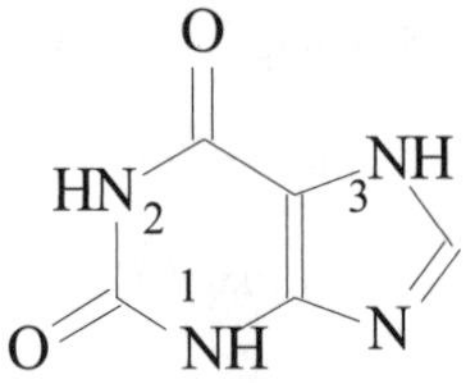

Figure 3.34.1 Xanthine. 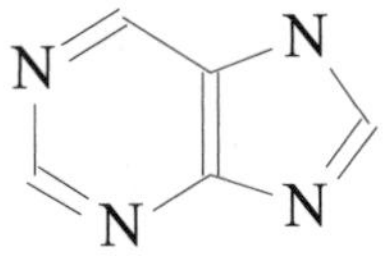   *Figure 3.34.2* Purine.

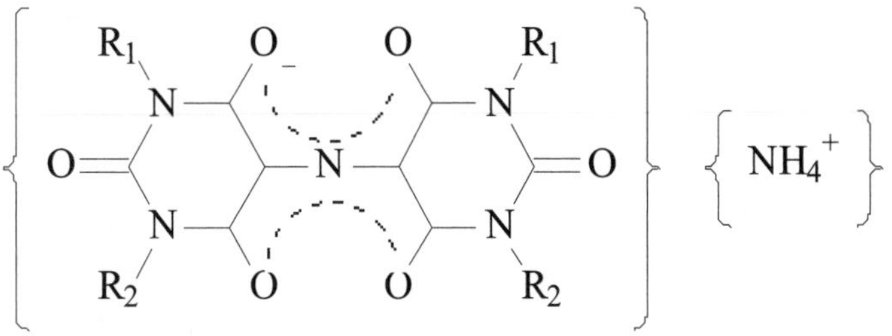

Figure 3.34.3 Murexoin ammonium salt.

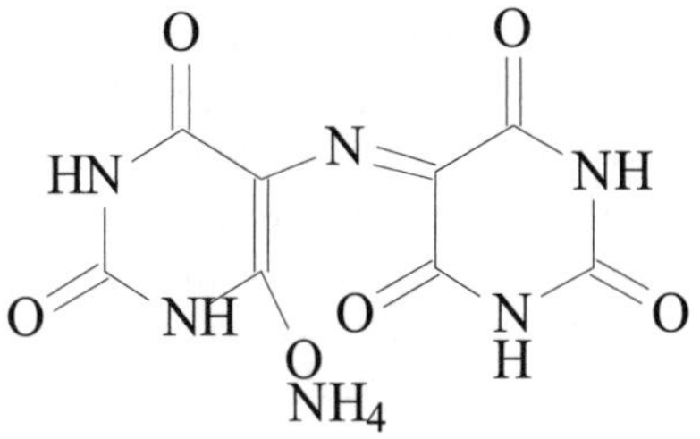

Figure 3.34.4 Murexide.

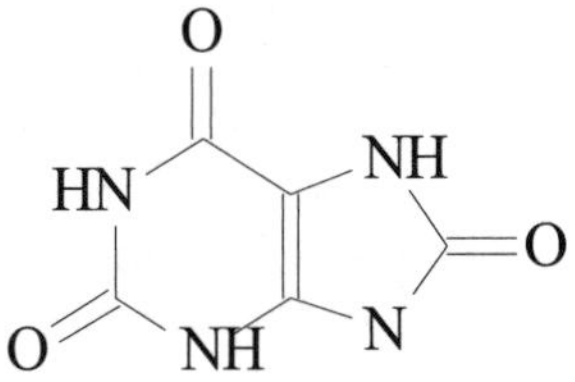

Figure 3.34.5 Uric acid.

The compound responsible for the purple color, or at least one of the compounds responsible, has been isolated. A larger number of compounds, which is believed to be responsible for the yellow reaction product of the first step of analysis, have also been isolated. But still, it is evident that a very complex chain of chemical reactions takes place during the procedure. The core of the reaction is, apparently, that the compound in question, through a number of steps, is oxidized into the intermediate alloxane that ultimately condenses to murexoin. The ammonium salt of murexoin is the principal contributor to the purple color of the final test solution. The murexide reaction of caffeine can therefore, extremely simplified, be written as above (Figure 3.34.6).

The N-methyl groups of murexoin stems from the N-methyl groups of caffeine, and will therefore, in the murexide product of the other xanthines, be absent or replaced by the alkyl groups of the table above. But this will not affect the appearance of the product.

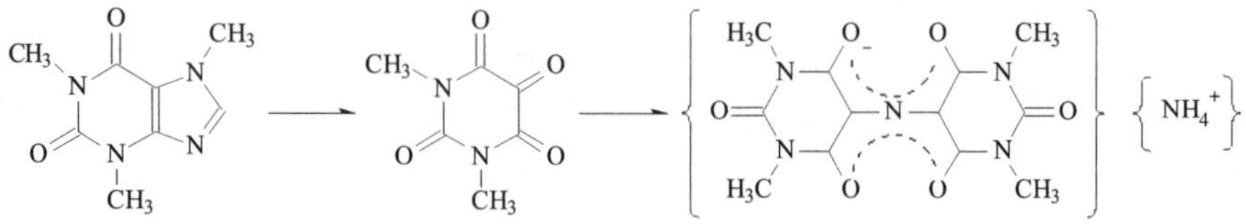

Figure 3.34.6 Murexide reaction of caffeine.

Different authors have used several experimental conditions. Most important is the use of different oxidizing agents and different concentrations of these. Changing the parameters appears to affect the reaction pathways and the intermediates that can be isolated at different steps of the procedure. Most likely this implies that a great deal of variance could be revealed if one would investigate in detail the reaction intermediates formed when the test as described by the European *Pharmacopoeia* is used on the different monographs. Possibly the reaction pathways are even sensitive toward relatively minor laboratory-to-laboratory and day-to-day differences in how the procedure is carried out. But it appears, however, that the general picture is that the end result achieved is not very sensitive toward reaction conditions. And the lack of an exact pathway, therefore, does not seem like a large obstacle.[1,2]

Whereas the exact reaction mechanisms of the test are not easily elucidated in detail, some knowledge has been gathered about the structural prerequisites a compound must meet before it can give a positive murexide reaction. It appears that some rules can be set up about the substituents on the 6-membered heterocyclic ring, including the nature of the substituents, which in the case of the xanthines includes a 5-membered heterocyclic ring.

The above structure (Figure 3.34.7) sums up the necessary elements. There must be a 6-membered hydropyrimidine structure. Closely related 5-membered rings like hydantoine (Figure 3.34.8) does not give a positive reaction. The X substituent on carbon number 2 must be oxygen or sulfur, so an imino group as in guanine (Figure 3.34.9) inhibits a positive reaction. The carbon numbers 4 and 5 must not have carbon substituents (at least not large groups) attached and consequently a negative reaction is given by propylthiouracil (Figure 3.34.10). The Y substituent at carbon number 6 must be oxygen or amino.[3]

It is important to remember that the reaction is not a group reaction for xanthines, but for a quite large and not easily defined group of compounds.

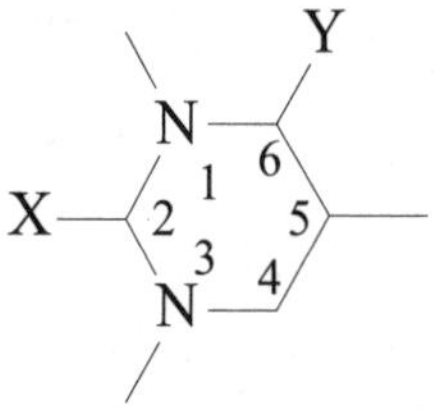

Figure 3.34.7 Structural requirements.

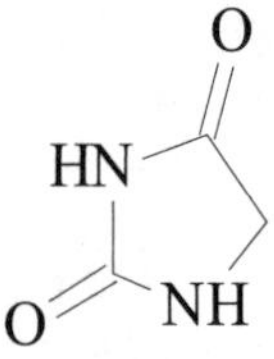

Figure 3.34.8 Hyantione.

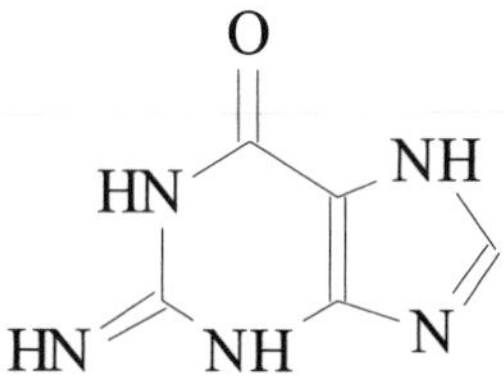

Figure 3.34.9 Guanine.

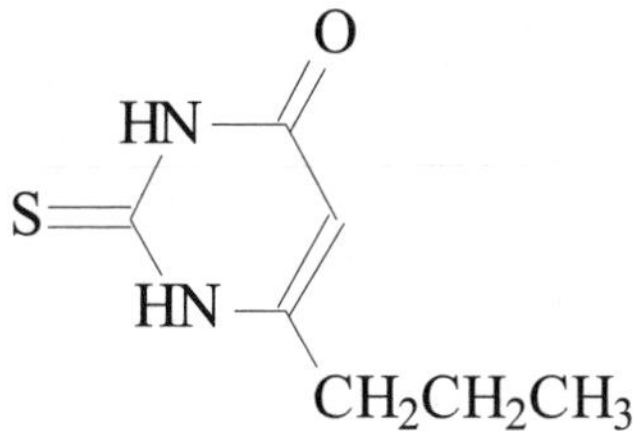

Figure 3.34.10 Propylthiouracil.

Drommond, nevertheless, has shown by testing a very large number of organic compounds that the chance of getting a positive reaction with a substance that does not meet the prerequisites defined above is not very big.[4]

References

1. Kozuka, H., Koyama, M. and Okitsu, T., Murexide reaction of caffeine with hydrogen peroxide and hydrochloric acid, *Chem. Pharmacal. Bull.*, 29, 1981, 433.
2. Koyama, M. and Kozuka, H., Murexide reaction of caffeine with hydrogen peroxide and hydrochloric acid. II, *J. Heterocyc Chem.*, 27, 1990, 667.
3. Bohle, F.J. and Auterhoff, H., Voraussetzungen der Murexidreaktion., *Archiv der Pharmazie*, 302, 1969, 604.
4. Drommond, F.G., Rex, E.H., and Poe, F., A study of the murexide test for caffeine and theobromine, *Analytica chimica acta*, 6, 1952, 112.

3.35 Zinc (European Pharmacopoeia 2.3.1)

The test identifies the presence of zinc, Zn^{2+}, in the substance to be examined. Reference to zinc is currently made in nine monographs. Besides those describing simple inorganic salts used as excipients, there are a few zinc salts of pharmaceutically active organic anions. The only possible ionic form of zinc is Zn^{2+}, and as the other d-block elements it has a strong tendency of forming complexes. It belongs to the class of cations that form amphoteric hydroxides. This, together with its unsoluble sulfide salt, is the basis of the identification. In solutions of soluble zinc salts, the zinc ion will be present as the hydrated ion $[Zn(H_2O)_6]^{2+}$. These solutions are mildly acidic since this ion hydrolysis relatively easy.

$$[Zn(H_2O)_6]^{2+} + H_2O \rightarrow [Zn(H_2O_5)OH]^{b+} + H_3O^+$$

In the majority of monographs, sample preparation is done simply by dissolving in water or in an acidified watery solution. In a few monographs of the organic salts, the anion is removed by ignition or liquid/liquid extraction.

In the first step, the 5 ml test solution is added to 0.2 ml strong *sodium hydroxide solution R*. A white gelatinous precipitate of zinc hydroxide is formed.

$$Zn^{2+} + 2OH^- \rightarrow Zn(OH)_2\downarrow$$

Since zinc hydroxide is amphoteric, it will redissolve, as further 2 ml of strong *sodium hydroxide solution R* is added in step two due to formation of the zinc hydro complex.

$$Zn(OH)_2\downarrow + 2OH^- \rightarrow [Zn(OH)_4]^{2-}$$

So far, the test would give a positive reaction with Be^{2+}, Pb^{2+}, Al^{3+}, Sn^{2+}, Sn^{4+}, and Sb^{3+}, since they all give white precipitate that redissolves in excess sodium hydroxide. The exact appearance of the respective precipitates differs somewhat but not to a degree that it would be noticed without performing a positive control. Step 3 in the procedure is to add 10 ml *ammonium chloride solution R*.

$$[Zn(OH)_4]^{2-} + 4NH_3 \rightarrow [Zn(NH_3)_4]^{2-} + 4OH^-$$

This causes the hydro complex to change into a soluble ammonia complex, the tetramminezincate(II) ion. If the observations in step one and two were caused by Be^{2+} or Al^{3+}, adding ammonium chloride would cause it to reprecipitate as hydroxides due to a lowering of the OH^- concentration (see 3.4. Aluminum). The last step is to add 0.1 ml *sodium sulfide solution R,* which causes the formation of a white flocculent precipitate of zinc sulfide.

$$[Zn(NH_3)_4]^{2-} + S^{2-} \rightarrow ZnS\downarrow + 4NH_3$$

A relatively high degree of selectivity is thereby achieved as the sulfide precipitates of Pb^{2+}, Sn^{2+}, Sn^{4+}, and Sb^{3+} are black, brown, yellow, and respectively, orange-red. This gives a very clearly negative test result, although, for example, Sn^{2+} most likely already has revealed itself by reprecipitating as a result of step two.

The possibility of a false positive reaction seems therefore unlikely or even exotic, but not impossible. But, when discriminating inorganic ions by their solubility as different salts and in different chemical environments, one should recognize that seemingly small variations in test conditions can change the selectivity and robustness of the test. Variations in sample concentration, pH and ionic strength of the test solution, and presence of a counter-ion can alter the situation from what is described above.[1-4]

References

1. Vogel, A.I. and Svehla, G., *Vogel's Qualitative Inorganic Analysis*, 6th ed., Longman Scientific & Technical, Essex, 1987, pp. 57, 91, 96, 98, 108, 127, 266.
2. Reimers, F., *The Basic Principles for Pharmacopoeial Tests*, Heinemann Medical Books, London, 1956, p. 23.
3. Brown, G.I., *Introduction to inorganic chemistry*, 1st ed., Longman, New York, 1974, pp. 359, 367.
4. Charlot, G., *Qualitative Inorganic Analysis. A New Physio-Chemical Approach*, Methuen & Co., London, 1954, p. 201.

Part II

Limit tests

chapter four

Precipitation in limit tests

The precipitation of an insoluble substance from a solvent is a complicated process and great effort has been put into understanding the nature of the steps involved. These steps are, regardless of the nature of the precipitate in question, nucleation and crystal growths. The response of precipitate formations to different sets of conditions, on the other hand, depends strongly on the type of substance being precipitated. It is therefore relevant to go through some of the principles of precipitation when investigating the nature of limit tests based on precipitation. Since the tests are dependent on the ability of the operator to make a visually reproducible precipitate in a test solution and a standard of a reasonably similar composition, this chapter will focus on how operating conditions will affect the formation and character of a precipitate.

The solubility of an ionic substance in a solvent is limited by its solubility product. The solubility product is the product of the molar concentrations of the individual salt-forming ions in a saturated solution of the salt in question. The solubility formula for the general salt A_aB_b is:

$$Ksp = [A^{z+}]^a \times [B^{z-}]^b$$

As different salts have very different solubility levels, usually the negative logarithm to the solubility product is used to make the vast span of solubility products encountered easier to manage.

$$pKsp = -\log (Ksp)$$

The first prerequisite for precipitation to take place is that the solubility of the actual substance in the solvent is exceeded. For an ionic salt this means that the ion product Q, is higher than the solubility product Ksp. The ion product is the product of the actual molar concentrations of the individual relevant ions, calculated as defined for the solubility product.

$$Q > Ksp$$

Normally, no precipitate is formed immediately if Q is only slightly above the solubility product. What initializes the formation of a precipitate is that a number of nuclei are created spontaneously in the solution. A nuclei is a very small entity that acts as a seed upon which crystal growth can be initiated and continue. It is often a tiny crystal of the precipitating substance itself, but it can also be that an impurity acts as a nuclei. In some precipitations, the nuclei needed to start the precipitation is very small. It has been revealed that the nuclei that initiates the precipitation of barium sulfate consists of only four barium ions and four sulfate ions.

The probability for nucleation to occur depends strongly on the degree of supersaturation of the solution. In the case of a very slight supersaturation, precipitation can take weeks or month to initiate, and at a certain degree of supersaturation nucleation and precipitation will happen immediately. This is called the critical supersaturation ratio. In the case of barium sulfate, a relatively high critical supersaturation ratio of about 32 has been determined. The supersaturation ratio is defined by the formula:[1]

$$\text{Supersaturation ratio} = (Q/Ks)^{1/2}$$

One should remember that the formula is used only to describe how fast after supersaturation is achieved one can expect precipitation, and not whether it will happen. The solubility product describes the supersaturated system thermodynamically and predicts that precipitation will take place. The supersaturation ratio describes the supersaturated system kinetically by prediction of when it will take place. Also one should remember the role of impurities in nucleation. In the original determination of the critical supersaturation ratio of barium sulfate, solvent and reagents of the lower purity were used, giving a value of 21 instead of 32.

Once nucleation has taken place, crystals will grow in an organized manner, determined both by the nature of the precipitate and also on the set conditions in the procedure. An important parameter in relation to limit tests is the size of the crystals. This also correlates with the degree of supersaturation in the mother phase, since crystal size is inversely proportional to the relative supersaturation as defined by the formula:

$$\text{Relative supersaturation} = (Q - S)/S$$

where Q is the molar concentration of the ionic compound before any precipitation takes place and S is the molar solubility of the precipitate when the system has come into equilibrium. Please notice that Q in this formula is not the ion product as it was in the preceding formulas.

Large crystals are therefore obtained when precipitating from a weakly supersaturated solution, and small crystals are obtained when precipitating

---

[1] This formula, as well as the formula for *K*sp and *Q*, is throughout the text simplified by using molar concentrations instead of activities.

from a strongly supersaturated solution. This is of great importance for several reasons. First, it is important since the visual appearance of a precipitate in a turbid solution correlates not only with the actual concentration of insoluble material suspended, but indeed also with the size of the crystals. Second, crystal size is important since some precipitate shows a variation in solubility with varying particle size.

A salt like barium sulfate has a solubility product of 1.1×10^{-10}, but particles with a size of 0.04 µm has a solubility about 1,000 times higher than for the coarse product. This is a phenomenon restricted to some compounds, and is normally seen for particles of a size less than 1 µm only. A salt like silver chloride does not show a significant particle size variation in solubility.

This means that a supersaturated solution of silver chloride will quickly form a high number of nuclei, even if Q is only moderately above the solubility product. The critical supersaturation ratio is low. In the case of a slightly supersaturated barium sulfate solution, nucleation will be much less pronounced because of the high solubility of the small particles, and this leads to fewer but larger crystals. This explains its high critical supersaturation ratio. In addition, a precipitate of the barium sulfate type has a tendency of exhibiting crystal growth on standing. In time molecules will redissolve from the highly soluble small particles and reappear in the precipitate as crystal growth on the larger particles. This will be seen only to a minor extent in a silver chloride–type precipitate, however its appearance can of course alter due to other reasons such as coagulation.

So where silver chloride–type precipitates tend to form colloidal suspensions despite the initial degree of supersaturation, the appearance of a barium sulfate precipitate is more dependent on the initial degree of supersaturation and on time of standing before evaluation.

There is some correlation between the hardness of the crystal and the chance that its shows a particle size dependent solubility. In general, hard precipitates, like barium sulfate, show particle size–dependent solubility, whereas soft precipitates like silver chloride do not.

In relation to limit tests, the general trend will therefore be: soft precipitates give a precipitate of a relative small particle size distribution regardless of the experimental procedures used and will not change in time. The particle size distribution of a hard precipitate depends on sample handling and might display crystal growth.

But other factors that are specific to the actual precipitate in question can influence the size of the particles obtained, and this is especially important in relation to the aim of obtaining a precipitate of similar appearance in the standard and the test respectively. Differences in pH and ionic strength between the two can give great differences, not only in particle size but even more pronounced in the macroscopic appearance of the precipitate. This again is largely determined by the precipitates affinity toward the water molecules and ions of the solvent. This is described in Chapter 1, "Precipitation in Identifications."

The strength of a colloidal suspension can be measured by nephelometric or turbidimetric method. These two methods are based on the fact that the particles of a suspension absorb and scatter a part of the light that passes through them. The principle in nephelometric calculations is to determine the amount of light scattered by the suspension by measuring the light intensity at an angle of 45° relative to the incident light. In turbidimetric determinations the amount of light absorbed by the suspension is calculated by measuring the transmitted light in the same angle as the incident light, much as in absorption spectrophotometry. Absorbance and scattering vary with particle size distribution of the crystal in suspension, as well as with the nature of the particles and their shape, but they are not proportional. Both the absorbance and scattering of a suspension increase with particle size, but the maximum absorbance is normally found at a larger particle size than the maximum scattering. The result of the nephelometric or turbidimetric determination is normally compared to that of a series a well-defined standards to produce a quantitative result.

In the *European Pharmacopoeias* limit tests, samples and standards are not compared using an apparatus as in nephelometric or turbidimetric determinations, but by visually comparing the turbidity or opalescence of the two. In principle though, despite the lack of instrumentation, the same concerns as described above have relevance. One can visually judge opalescence in the way for example described in 6.4. Chloride and the general pharmacopoeia method laterally against a black background using test tubes with a flat base, or one could judge the opalescence looking through the suspension against a lid up background as most operators would in 6.8. Heavy metals. The first method resembles a nephelometric determination, the other one a turbidimetric determination. Which one is the most correct depends naturally on the color of the precipitate, and the relative opalescence/turbidity should not depend on the way it is judged. An important exception is, however, for example, if the particle size distribution is different in the standard and sample respectively, due to some of the factors describe above. Another problem could arise when the opalescence of a white precipitate in a standard and sample is to be compared. If the sample, apart from the opalescence given by the precipitate, unintentionally should have a dark coloration arising from contamination or a misconducted sample preparation, one could have a result that depends on the way they are compared. If judged against a black background, the relative opalescence of the standard and sample most likely corresponds to the strength of the two suspensions. But if viewed against a lighted background, the suspended material will appear dark and the dark coloration of the sample would wrongfully be superimposed on this, leaving a wrongful impression of the sample purity.

chapter five

Color reactions in limit tests

Visually comparing the color intensity of a sample with a standard of known concentration, after reacting both with a selective reagent, is a very old analytical principle. In the second half of the nineteenth century, it was discovered that several (colorless) alkaloids and other pharmaceutical substances produced characteristic colors when treated with certain inorganic reagents. These reactions made possible semiquantitative determinations of the substances, but there was most often no knowledge about the structure of the colored product formed or the reaction mechanism involved in the reaction.

During the first two decades of the 20th century, simple optical instruments were developed with which the analyst could visually compare the color intensity of a sample with a standard. These photometers with visual light source used colored glass filters to select a wavelength interval suitable for the substance to be analyzed, and their applicability was limited to colored compounds or colored derivatives.

This gave an extensive search into developing novel selective color reagents capable of lifting a low wavelength absorption spectra of a given analyt up in the visible area and into the measuring range of the available apparatuses. A large fraction of the analytical chemistry literature published was concerned with aspects of this type of analysis, and many official standards based their official analysis on this methodology, including the pharmacopoeias. In the present day laboratory chromatographic and spectroscopic methods have widely replaced the more labor intensive color reactions, and this clearly shows in the topics dealt with in contemporary scientific literature.

Tests based on visual comparison of a standard and a sample were named colorimetric analyses, whereas photometric analysis make used of a photometer. Spectrophotometric analysis is based on spectrophotometers with incorporated monochromoters and detector.

Of prime interest, when considering general aspects of color reactions used in limit tests, is the specificity of the chemical reaction between the color reagent and the analyt in question, as well as factors (besides analyt concentration) contributing to the color intensity of the sample and standard respectively. The chemical selectively relies on the actual color reagent used and the analyt being examined and will be discussed in the individual chapters, but a few general remarks can be given on the subject of color intensity variations.

Compounds containing an acid or base group will most likely have a pH dependent UV-visible absorption spectra. If there are large changes in the visible region it will be acknowledgeable as a change in color, a property forming the basis of pH indicators. Only compounds where the acid or base group contributes with their n or pi electrons to the conjugated electron systems of the molecule will show marked changes in absorption spectrum and in appearance.

An obvious example is the pH indicator methyl red, which is red in its acid form (Figure 5.1) and change to its yellow basic form (Figure 5.2) at a pH near 5.1.

This marked change in absorption behavior is equally evident in the absorption spectras measured at pH on both sides of the pKa of the indicator (Figure 5.3).

As many of the color reagents used by the pharmacopoeia do contain an acid or base group even after reaction with the analyt, this behavior is often encountered in test and standard solutions. It means that a difference in pH and ionic strength between the standard and sample solution can give a difference in color intensity, if not even a difference in color, even with equal concentrations of the colored product.

In this discussion, it is important to remember the less obvious acid and bases, for example enolisable ketones and aldehydes. These compounds will be found partly in their enol resonans form (Figure 5.4).

In alkaline conditions, the equilibrium is likewise but with the enolate ion (Figure 5.5).

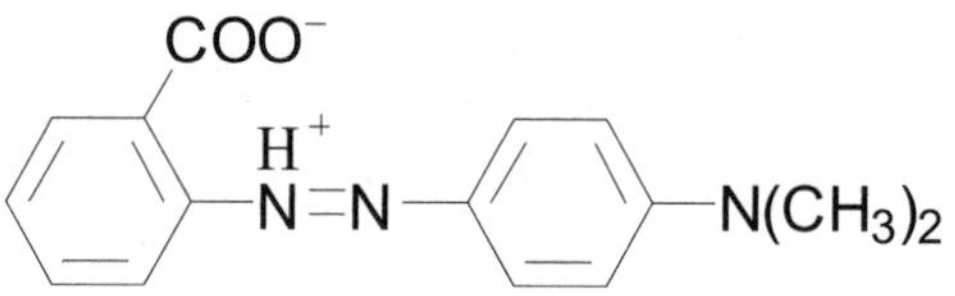

Figure 5.1 Methyl red, red acid form.

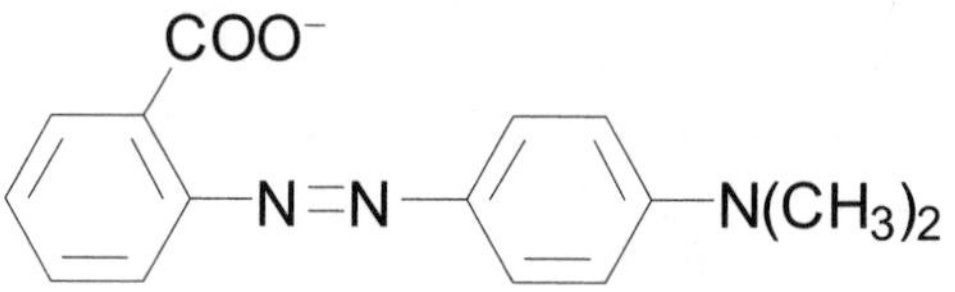

Figure 5.2 Methyl red, yellow basic form.

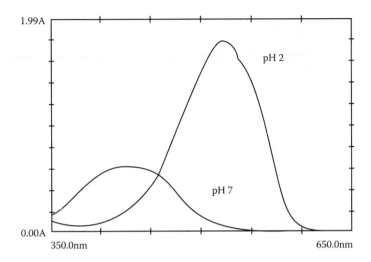

Figure 5.3 Spectrum of methyl red at pH 2 and 7.

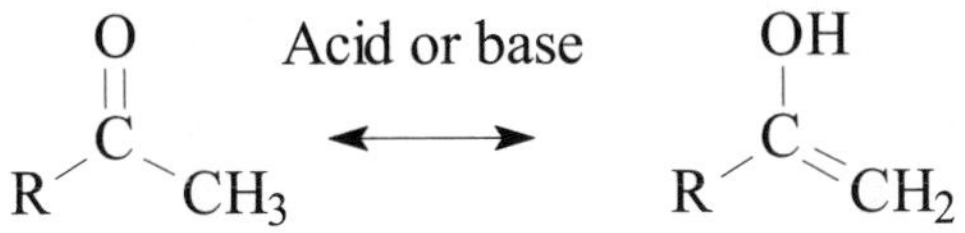

Figure 5.4 Ketol-enol equilibrium.

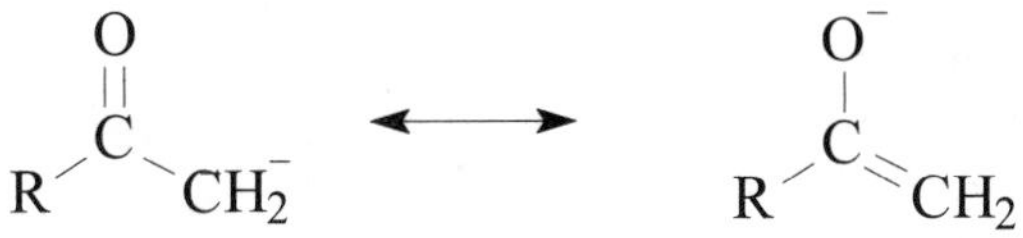

Figure 5.5 Resonance.

Even though the colored compound in question does not contain a carboxylic acid group or a nitrogen base, but does contain, for example, an oxo group, it might have a proteolytic ability and show a pH dependent absorbance or appearance.

If the acid or base group of a compound participates in a salt formation, the absorption spectra and in some cases the color intensity or appearance of the parent compound will be affected. Same behavior is seen when a compound participates in a complex formation with a metal, and this is widely used in determining metal ions with the aid of organic color reagents. The same effect can, however, also be the cause of unintended interference in cases were a test solution or reagent solution in a color reaction is contaminated with metal ions.

Complex forming metals are usually found in the *d* block of the periodic table as listed in Table 5.1 (or the rare *f* block) shown in Table 5.1. They form

Table 5.1 D Block Elements

H																	He
Li	Be											B	C	N	O	F	Ne
Na	Mg											Al	Si	P	S	Cl	Ar
K	Ca	Sc	Ti	V	Cr	Mn	Fe	Co	Ni	Cu	Zn	Ga	Ge	As	Se	Br	Kr
Rb	Sr	Y	Zr	Nb	Mo	Tc	Ru	Rh	Pd	Ag	Cd	In	Sn	Sb	Te	I	Xe
Cs	Ba	La	Hf	Ta	W	Re	Os	Ir	Pt	Au	Hg	Tl	Pb	Bi	Po	At	Rn
Fr	Ra	Ac	Unq	Unp	Unh	Uns	Uno	Unn									

complexes due to their ability to act as Lewis bases, and the tendency of the complexes to be colored is because their *d* and *f* electrons are easily excitable.

A Lewis acid is an electron pair acceptor and a Lewis base is an electron pair donator. A complex is formed when a Lewis acid and a Lewis base share a pair of electrons and form a bond. When an organic compound is used as a ligand in a complex formation with a metal it acts as a Lewis acid and the combination in many cases produces real acids. This means that the metal Lewis bases compete with the acidic "loose protons" for the sharing electron and will be replaced by the proton at pH values that are low compared to the pKa of the ligand in question. The metals of the *d* block, on the other hand, are known for forming insoluble hydroxides or oxides at neutrality or high pH. So a complex is often stable only inside the pH interval between ligand deprotonation and metal hydroxide precipitation.

The intensity of a condensated color product is similarly pH dependent, but through a somewhat different mechanism. Here two conditions have to be fulfilled. The analyt and the reagent should be mixed in a test solution with a pH facilitating formation of the colored condensate, and when this product is formed it is desirable to shift the test solution pH toward a value giving the optimum color intensity the best stability achievable. It is very likely that the pH optimum of the reactions and of the product are alike, and a difference between the standard and test solution pH can therefore give rise to color intensity differences through two different mechanisms.

Another general remark one could make regarding colored condensation products is that the chemical reactions affording them are in many cases reversible and that an equilibrium therefore is established. This equilibrium might not be shifted entirely to the right. A quantitative reaction is by no means a requisite for the analysis to be valuable, as long as the equilibrium constant in the test solution and standard are alike, giving a similar yield of colored product in both solutions. This again brings focus on ensuring that the chemical environment in the standard and test are alike since many factors influence a chemical equilibrium, including pH, ionic strength reagent concentration, and temperature. The last parameter affects both the position of the equilibrium and the speed with which the equilibrium is reached. Many chemical reactions have reaction kinetics in which a 10°C change in temperature affords a two- to threefold increase in reaction speed.

The absorption spectra or even appearance of some colored compounds varies with the nature and concentration of the solvent in which they are dissolved. Polar solvents are capable of giving weak interactions with the electrons of a molecule's conjugated system, whereas nonpolar solvents are not. Much effort has been given into elaborating systems capable of predicting the shift in absorbance maximum and intensity caused by a given change in solvent, and with some success too. An example of solvents affecting visual appearance is found in 3.19. Iodides.

Much of the literature dealing with the individual color reactions found in the pharmacopoeia describe methods for quantitative spectrophotometric determinations, whereas in the pharmacopoeia they most often are simple colorimetric tests. In these cases, it is valuable to correlate the visual color of a complex with its absorbance maximum. This is tabulated below in Table 5.2 correlation, where the different maximum absorbance wavelength intervals are correlated with the expected color of the compound, hue (transmitted), and with the color of the absorbed wavelengths, complementary hue.

Table 5.2 Transmitted and Complementary Hue

Wavelength (nm)	Hue (transmitted)	Complementary hue
400–435	Violet	Yellowish-green
435–480	Blue	Yellow
480–490	Greenish-blue	Orange
490–500	Bluish-green	Red
500–560	Green	Purple
560–580	Yellowish-green	Violet
580–595	Yellow	Blue
595–610	Orange	Greenish-blue
610–750	Red	Bluish-green

chapter six

Tests

The *European Pharmacopoeias* general Chapter 2.4 Limit tests include 27 monographs, each describing a test developed to assure that any unintended content of a well-defined element or compound is below some critical limit. The reason for limiting the element or compound in question varies from monograph to monograph. Some monographs reveal a residual content of toxic substances, like arsenic and heavy metals, and the reason for limiting their presence is obvious. Others, however, cover clearly harmless elements like chlorides or the alkali metals. In those cases, the reason for limiting their presence is one of general product quality. A well-produced, active pharmaceutical ingredient or excipient need not have a residual content of, for example, inorganic salts used during synthesis. When present, however, they reveal a product of low general quality.

The typical *European Pharmacopoeia* raw material monograph is made up of several tests which determine the purity of the substance. Some of these tests can be used only on the specific material covered by the monograph and the analytical procedures of the tests are described in full. The most widely used tests, on the contrary, are described in the general method part of the pharmacopeia including the 2.4. Limit tests. Using the same test on a variety of different materials will not only put demands on the performance of the tests but also put demands on the skills and knowledge of the operator doing the testing. The aim of the chapter is provide the reader with some insight into the principles of the individual tests and function as the starting point for further investigations.

Out of the 27 tests of the Chapter 2.4 Limit tests, 10 of the tests described require sophisticated analytical equipment such as gas chromatographs whereas the remaining tests rely on bench top chemistry. Because a wealth of contemporary literature involves the use of instrumented analytical chemistry, this chapter focuses on the older wet-chemistry-based methods only.

6.1 Ammonium (European Pharmacopoeia 2.4.1)

The test ensures that the substance to be examined has a total content of ammonium (NH_4^+), which is below the limit defined in the individual monograph. Ammonium is referenced in less than 50 monographs and of these amino acids and amino acid derivatives constitute about half of them. It is explicitly stated in Ammonium that, unless otherwise prescribed, method (a) is to be used. Method (b) is used almost exclusively for testing amino acids and derivatives, since method (a) in many cases does not work with those. The majority of monographs referencing method (a) describe inorganic salts where ammonium is potentially present as a reagent residue. In the case of amino acid ammonium can stem from ammonium buffers used in chromatographic purification. Ammonium is not toxic and its presence in pharmaceutical raw material is limited due to a quality rather than a safety concern. The amount of residual ammonium allowed varies from 10 ppm to 0.2%. Ammonium is almost identical in size to potassium, and therefore has properties almost identical to this cation. It forms water-soluble salts with almost all inorganic anions. The two tests of the pharmacopoeia are based on the Nesslers reagent that gives a precipitate of potassium tetraiodomercurate(II), and on a redox reaction facilitated by the presence of ammonia.

Method (a)

In 1856, J. Nessler was the first to propose an alkaline solution of mercury(II) iodide in potassium iodide as a reagent for the colorimetric determination of ammonia. It has been widely used especially in determining small amounts of ammonia in simple solutions as, for example, drinking water. It is, unfortunately, due to its limited selectivity, feasible only if all other metals except sodium and potassium are absent.

The pharmacopoeia reagent is a relatively weak solution prepared of *potassium iodide R* and *mercuric iodide R* which is further diluted immediately before use. In this it is different from the reagent described in classical Nessler's reagent, which is a saturated filtered solution. The final reagent is a alkaline solution of potassium tetraiodomercurate(II), $K_2[HgI_4]$.

The prescribed quantity of the substance to be examined is dissolved in 14 ml of *water R* in a test tube, made alkaline if necessary by the addition of *dilute sodium hydroxide solution R* and diluted to 15 ml with *water R*. To this solution 0.3 ml of *alkaline potassium tetraiodomercurate R* is added. A standard is prepared by mixing 10 ml of *ammonium standard solution (1 ppm NH₄) R* with 5 ml of *water R* and 0.3 ml of *alkaline potassium tetraiodomercurate R*. Stopper the test tubes. After 5 min, any yellow color in the test solution is not more intense than that in the standard.

In the conditions applied, ammonia will give a yellow color due to the formation of the high molecular structure show below.[1]

$$[NHg_2I]_x$$

It is really not a colored complex but a colloidal precipitate and with increasing concentration it will give first a red brown coloration and eventually a red brown precipitate. The intensity of the color formed correlates to the speed with which the test solution and the reagent are mixed. Contrary to what is most often seen in precipitation a fast mixing results in a slow formation of the color, which is the reason for evaluating after 5 min. A satisfactory reproducibility can, however, be obtained if reagent addition and mixing is done in a standardized fashion among standard and test solution.

The tendency of the reaction product to form not just a colloidal but a real turbid precipitate has been claimed to compromise the standard test solution comparison and destroy reproducibility. If a very small amount of crystals is present in the test tube it might cause the determination to give an unclear appearance through seeding, perhaps caused by a film of mercury present as residue from earlier determinations.[2] The risk of this is greater in high salt concentration solutions.

As previously mentioned, a high number of substances can be expected to interfere with the determination through altering the sensitivity, by precipitating with the reagent, or through incompatibility with the chemical environment of the test conditions. The most obvious example of the latter is that substances, for example, many cations but also organic compounds, that are insoluble at the high pH of the test will disturb the determination by precipitating. Halogens will lower the sensitivity of the test, an effect that increases through the series chloride, bromide, and iodide. Many organic substances affect the sensitivity too. It is affected by test solution pH in the way that sensitivity is lowered with lowered pH. Interference can be expected from organic amines, aldehydes, ketones, alcohols, chloramines, magnesium, manganese, iron, and sulfide. As a result of these limitations of the determination, ammonia is in many variants of the test first separated from the sample matrix through distillation.[3]

Due to the sensitivity of the reagent, solutions must be made with distilled water to minimize the content of ammonium and possibly other ions.

Method (b)

This method, which in the pharmacopoeia primarily has found use for limiting ammonium in amino acids, was originally used to detect and determine silver or manganese ions. In the presence of alkali ionic silver will be reduced to metallic silver by manganese(II).

$$Mn^{2+} + 2Ag^+ + 4OH^- \rightarrow MnO_2 + 2Ag^0 + 2H_2O$$

The silver and manganese dioxide thereby produced a gray to black coloration/precipitate that can be evaluated. The reaction will move to the right only in alkaline conditions, and this is used as the pharmacopoeial ammonium test.

In a 25 ml jar fitted with a cap, the prescribed quantity of the finely powdered substance to be examined is placed and then dissolved or suspended in 1 ml of *water R* and 0.30 g of *heavy magnesium oxide R*. Close immediately after placing a piece of *silver manganese paper R* 5 mm square, wetted with a few drops of *water R*, under the polyethylene cap. The jar is swirled, avoiding projections of liquid, and allowed to stand at 40°C for 30 min. If the silver manganese paper shows a gray color, it is not more intense than that of a standard prepared at the same time and in the same manner using the prescribed volume of *ammonium standard solution (1 ppm NH_4) R*, 1 ml of *water R*, and 0.30 g of *heavy magnesium oxide R*.

Any ammonium present in the substance to be examined will, by the addition of the alkaline magnesium oxide, be liberated as ammonium.

$$NH_4^+ + OH^- \rightarrow NH_3\uparrow + H_2O$$

But since magnesium oxide is a base of only moderate strength it will not be able to liberate ammonia from ammonium salts, and is not capable of hydrolyzing amides and other nitrogen bearing organic molecules, as would, for example, the alkali metal hydroxides.[4] Through the heat applied in the test ammonia will evaporate and afterward be caught by the silver manganese paper where the following reaction will take place:

$$4NH_3 + Mn^{2+} + 2Ag^+ + 3H_2O \rightarrow MnO(OH)_2\downarrow + 2Ag\downarrow + 4NH_4^+$$

A number of metals are capable of giving a reaction similar to the one seen between manganese(II) and silver, but this is of no relevance since they need to be volatile to be brought from the sample solution to the silver manganese paper. But any substance that is sufficiently alkaline to enable the manganese silver reaction and volatile enough to evaporate from the sample solution under the conditions set could give a positive reaction in the test.[5]

References

1. Rüdorff, W. and Brodersen, K., Die struktur der millonschen base und einiger ihrer salze, *Zeitschrift für anorganische und allgemeine chemie*, 274, 1953, 323.
2. May, C.E. and Ross, H.P., The nesslerization of ammonia solution, *J. Am. Chem. Soc.*, 43, 1921, 2574.
3. Kodama, K., *Methods of Quantitative Inorganic Analysis*, Interscience Publishers, New York, 1963, p. 462.
4. Feigl, F., *Tüpfelanalyse*, Band 1 *Anorganisher teil*, Vierte deutsche auflage, Akademishe Verlagsgesellschaft, Frankfurt am Main, 1960, p. 244.
5. Feigl, F., *Qualitative Analysis by Spot Tests*, 3rd ed., Elsevier Publishing Company, New York, 1947, p. 184.

6.2 Arsenic (European Pharmacopoeia 2.4.2)

The test ensures that the substance to be examined has a total content of arsenic(III) or arsenic(V), which is below the limit defined in the individual monograph. Some of the most commonly used technical reagents, hydrochloric acid and sulfuric acid, were earlier in risk of being arsenic contaminated notably, thus, due to their widespread use, a relatively high number of monographs reference arsenic.

Arsenic is in the periodic table placed close to the border between the metals and the nonmetals, and it partly has the properties of both. This means that it can be found either as an ionic species having a valence of +3 or +5, or covalently bound in, for example, an organic structure. As with its fellow members in *p*-block group V, nitrogen and phosphor, it has a great tendency of forming oxo and oxo acid compounds. Arsenic(III) is, in analogy to phosphor, in an alkaline solution found as the oxo anion AsO_3^{3-} and in acidic solutions as the acid H_3AsO_3. Arsenic(III) is in the same way found as H_3AsO_4 or AsO_4^{3-} according to pH. Ammonium and alkali salts of arsenic(III) and arsenic(V) are water soluble, whereas the rest are insoluble. Inorganic arsenic compounds are confirmed human carcinogens, trivalents being more poisonous than pentavalents. Organic compounds are less toxic.

Method (a)

Determinations of very small amounts of arsenic, such as 0.1 mg or less, were traditionally done with the Marsh test or the Gutzeit test invented in the late nineteenth century. Method (a) of the European *Pharmacopoeia* can be considered a modified Gutzeit test, but whereas this originally used silver nitrate or mercury(II)chloride to detect any arsenic present, method (a) uses mercury(II)bromide.

Into the lower tube insert 50 mg to 60 mg of *lead acetate cotton R*, loosely packed, or a small plug of cotton and a rolled piece of *lead acetate paper R* weighing 50 mg to 60 mg. Between the flat surfaces of the tubes place a disc or a small square of *mercuric bromide paper R* large enough to cover the orifice of the tube (15 mm × 15 mm).

In the conical flask the prescribed quantity of the substance to be examined is dissolved in 25 ml of *water R*, or in the case of a solution, the prescribed volume is adjusted to 25 ml with *water R*. 15 ml of *hydrochloric acid R*, 0.1 ml of *stannous chloride R*, and 5 ml of *potassium iodide solution R* are added. The solution is allowed to stand for 15 min and 5 g of *activated zinc R* is then introduced. The two parts of the apparatus are assembled immediately and the flask is immersed in a bath of water at a temperature such that a uniform evolution of gas is maintained. A standard is prepared in the same manner, using 1 ml of *arsenic standard solution (1 ppm As) R* diluted to 25 ml with *water R*. After not less than 2 h the stain produced on the mercuric bromide paper in the test is not more intense than that in the standard.

Under the conditions given arsenic(III) will be volatilized by nascent hydrogen to the gas arsinic, AsH_3. Any arsenic(V) present is first reduced by the iodide to arsenic(III) which is then volatilized.

$$As^{3+} + 3Zn + 3H^+ \rightarrow AsH_3\uparrow + 3Zn^{2+}$$

Hydrogen gas is formed concurrently and helps sweep the arsin.

$$2H^+ + Zn \rightarrow H_2\uparrow + Zn^{2+}$$

Stannous chloride is added since its presence facilitates the arsin formation. The reaction starts upon addition of the zinc and will often evolve very rapidly so that most of the arsenic present will be formed within the first few minutes. This especially is true if the zinc is a fine powder. For this reason the zinc used for the standard and the sample must be of similar quality in terms of specific surface and activity.[1] Other reaction conditions, of course, must also be standardized. The zinc is activated, which means that a small layer of platinum covers the zinc metal surface since this greatly increases the speed of the proton reduction through catalysis. But this activation is by some claimed to give more hydrogen evolution without necessarily giving more arsin. Nitric acid, chlorine, bromine, iodine, and compounds that give hydrogen sulfide must be absent, and mercury, platinum, silver, palladium, nickel, cobalt, and large amounts of copper and their salts are undesirable. Antimony interferes only if in excess.

Any hydrogen sulfide evolved in the conical flask is removed by the lead acetate paper since lead precipitates with the sulfide ion. Any escaping hydrogen sulfide will turn mercuric bromide gray-black. Phosphine, PH_3, and stibine, SbH_3, are also removed in the same way.

$$Pb^{2+} + H_2S \rightarrow PbS\downarrow + 2H^+$$

The product formed and the reaction pathway involved in the coloration of the mercuric bromide reagent and arsin are apparently not fully elucidated, but the sequence of reactions shown below has been suggested.

$$AsH_3 + HgBr_2 \rightarrow AsH_2(HgBr)$$

$$AsH_2(HgBr) + HgBr_2 \rightarrow AsH(HgBr)_2$$

$$AsH(HgBr)_2 + HgBr_2 \rightarrow As(HgBr)_3$$

$$As(HgBr)_3 + AsH_3 \rightarrow As_2Hg_3$$

Light and moisture fade the spot formed and in some procedures the spot is developed in a solution of an iodine salt to prevent the fading.[2-3]

Method (b)

The methodology used in method (b) is a modification of the one used in the arsenic identification test of the *European Pharmacopoeia*, 2.4.2. Arsenic.

The prescribed quantity of the substance to be examined is introduced into a test tube containing 4 ml of *hydrochloric acid R* and about 5 mg of *potassium iodide R* and 3 ml of *hypophosphorous reagent R* are added. The mixture is heated on a water-bath for 15 min, with occasional shaking. A standard is prepared in the same manner, using 0.5 ml of *arsenic standard solution (1 ppm As) R*. After heating on the water-bath, any color in the test solution is not more intense than that in the standard.

Any arsenic content present in the substance to be examined as either arsenic(III) or arsenic(V) will react with the hydrochloric content to form arsenic(III) chloride, in the case of arsenic(III) directly as shown below, or in the case of arsenic(V) after reduction to arsenic(III).

$$H_3AsO_3 + 3HCl \rightarrow AsCl_3 + 3H_2O$$

The arsenic(III) chloride will be reduced by hypophosphorous acid to amorph colloidal metallic arsenic, which appears more as a coloration than as a precipitate.

$$2AsCl_3 + 3H_3PO_2 + 3H_2O \rightarrow 2As\downarrow + 3H_3PO_3 + 6HCl$$

The precipitation differs quite a bit from the majority of the other turbidimetric tests where the addition of a reagent creates a situation of supersaturation. These tests are often rather sensitive to variation in parameters such as temperature, speed of reagent mixing, and agitation. The metallic arsenic formation is insensitive to such variations since the formation of the identifying compound is not controlled by the degree of supersaturation achieved by mixing but a chemical reaction controlled by the heat applied. This is a rather stable parameter, but the hydrochloric acid concentration affects the speed of reduction as well. If the concentration is lower than intended, reduction is inhibited and if it is to high, the risk is that the volatile arsenic(III) chloride is lost through evaporation. Excessive and prolonged heat can have the same effect. Also the temperature of the test solution when evaluated is without consequence.

In 1907, it was suggested by Bougault to have potassium iodide added to the test solution. The aim was both to enhance sensitivity and to speed up the speed of reduction, even at ambient temperature. The amount added in the pharmacopoeia test is very small compared to what was normally recommended for these purposes, so maybe the reason was really to use the masking properties of the iodide. If the test solution contains iron(III) ions they will give a yellow appearance. The iodide added will reduce these to iron(II), which has a much less yellow appearance. This reduction will be quantitative even if iron(III) is in excess to the iodide since the hypophosphorous acid continuously will reform the iodine to iodide. This had some importance especially earlier since iron was a quite common impurity in reagents and since even a very small amount of iron(III) gives quite a strong coloration.

The iodide addition has, however, a drawback in that it can reduce sulfates to sulfur, thereby giving yellow precipitation and destroying the test. The harsh reaction conditions of the test can also give another effect in that the chloride might precipitate with an alkali counterion. But the precipitate might not be a problem. Also the test cannot be used for testing substances susceptible to reduction, but since hypophosphorous acid is a relatively weak reagent, many organic substances can take it, but many will be charred with concentrated hydrochloric acid.

References

1. Goldstone, N.I., A source of error in the Gutzeit method for arsenic, *Ind. Eng. Chem.*, 18, 1946, 797.
2. Hildebrand, W.F. et al., *Applied Inorganic Analysis*, John Wiley and Sons, New York, 1953, p. 269.
3. Kodama, K., *Methods of Quantitative Inorganic Analysis*, Interscience Publishers, New York, 1963, p. 194.
4. Lachele, C.E., Rapid method for determination of small amounts of arsenic, *Ind. Eng. Chem.*, 6, 1934, 256.
5. Reimers, F. and Gottlieb, K.R., *Contributions from the Danish pharmacopoeia commission*, Vol. 1, *Limit Tests for Impurities*, Ejnar munksgaard, København, 1946, pp. 19, 69.

6.3 Calcium (European Pharmacopoeia 2.4.3)

The test ensures that the substance to be examined has a total content of calcium (Ca^{2+}), which is below the limit defined in the individual monograph. The calcium is like the rest of the alkaline-earth metals — a divalent cation. It forms soluble halogen and nitrate salts, whereas the carbonate, sulfate, phosphate, and oxalate are insoluble. Calcium is not toxic but on the contrary an essential mineral in the human diet. So, the main reason for limiting its presence in substances used in pharmaceuticals is that it could be used as a marker for insufficient reagent cleanup during synthesis of the substance. Some carboxylic acid, like lactic acid, is first precipitated as calcium salts upon which purification steps are then performed. After this the carboxylic acid can be reprecipitated as whatever salt needed. At the present reference to calcium is given in more than 100 monographs.

The test compares the turbidity caused by precipitation of the calcium present in the substance to be examined with oxalic acid (Figure 6.3.1) to the

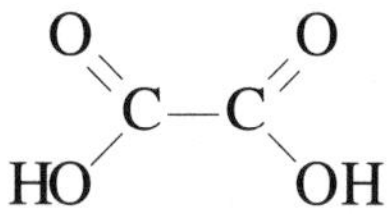

Figure 6.3.1 Oxalic acid.

turbidity formed under similar conditions in a standard of known calcium concentration.

Calcium oxalate is a hard crystalline precipitate and is therefore similar to barium sulfate with respect to the process of precipitation. This means that the crystal size distribution of the precipitated form varies a lot with precipitation parameters such as sample solution temperature, degree of super saturation, and time allowed from precipitation to evaluation of turbidity. For a more thorough discussion of this phenomenon, please see Chapter 4, "Precipitation in Limit Tests."

It is stated in the start of the text of the pharmacopoeia that all solutions used should be prepared with *distilled water R*. This is a necessity since tap water contains a relatively high amount of calcium ions especially if it is ground water. In addition to this, tap water might contain other ions that could influence the degree of opalescence obtained.

A test solution is prepared by mixing 0.2 ml of *alcoholic calcium standard solution (100 ppm Ca) R* and 1 ml of *ammonium oxalate solution R*. After 1 min, a mixture of 1 ml of dilute *acetic acid R* and 15 ml of a solution containing the prescribed quantity of the substance to be examined is added and the solution is shaken.

A standard is prepared in the same manner using a mixture of 10 ml of *aqueous calcium standard solution (10 ppm Ca) R*, 1 ml of dilute *acetic acid R*, and 5 ml of *distilled water R*. After 15 min, any opalescence in the test solution should not be more intense than that in the standard.

Calcium gives, with oxalic acid under these conditions, a precipitate of calcium oxalate.

$$Ca^{2+} + (COO)_2^{2-} \rightarrow Ca(COO)_2\downarrow$$

Obtaining a reproducible precipitate is not very easy since small calcium oxalate crystals have a solubility product that differs dramatically from the solubility product of larger crystals. This means that a highly supersaturated solution of calcium oxalate gives a precipitate with very fine crystals, whereas an only slightly supersaturated solution gives a precipitate of significantly larger crystal size distribution. It also means that a calcium oxalate will show crystal growth because the larger less soluble crystals will grow at the expense of the smaller more soluble crystals. So even when using well-defined operations one can easily obtain a sample with an opalescence twice that obtained when using slightly different operations, and some of the steps in the precipitation, as they are described in the pharmacopoeia, are intended to limit this lack of reproducibility.

In the first step of the analysis, a crystal seed is precipitated by mixing a small volume of *alcoholic calcium standard solution (100 ppm Ca) R* with an execs of *ammonium oxalate solution R*. The precipitation is carried in a high concentration compared to the sample and in ammonia since this environment also facilitates precipitation. The precipitation is performed in alcohol since calcium oxalate is less soluble in this and therefore more likely to precipitate faster with small crystals. The amount of calcium precipitated in this step is only one-fifth of the calcium added as residue in the substance to be examined or as standard. And when this second precipitation is carried out on an already existing crystal seed it gives a higher reproducibility than if carried out without seeding.

The second precipitation is carried out in an acetic acid environment instead of ammonia as used in the first precipitation. This is done to enhance selectivity of the test since nearly all metals, except alkali metals and iron(II), form insoluble or sparingly soluble salts with oxalic acid at neutral conditions. For example, barium and a number of other cations will give a precipitate which is soluble in dilute acetic acid. Even precipitated calcium oxalate will redissolve if a stronger acid is added as the concentration of the anionic species of oxalic acid declines in accordance with its pKa values of 2.2 and 3.6. The lack of selectivity of the test should be remembered in case of an unexpected result.

Despite the efforts put into enhancing the reproducibility of the test it is still advised to perform the test with the greatest possible care. The speed with which calcium and oxalate solution is mixed plays a great role for the appearance of the precipitate obtained, and the step should be performed exactly alike in standard and sample solutions. Deviations from the time intervals dictated in the test might give unexpected results, especially if deviation varies among the standard and sample. All test solution must be shaken in the same way. Temperature of the sample solution and the standard solution must be similar since the solubility of calcium oxalate declines quite steeply with increasing temperature. This implies that opalescence increases with increasing temperature.[1]

A very thorough investigation into the effect of operational parameters onto the result obtained in the limit tests of calcium and sulfates has been published by Zimmermann et al.[2]

The presence of other salts lowers the sensitivity of the test.[3] Sodium chloride, sodium sulfate, and potassium reduce sensitivity to about half of what it is without, but tartrate and citrate lowers it much more by forming complexes with calcium. Aluminum forms a complex with oxalate and will thereby replenish the reagent if present in sufficient concentration.[4]

References

1. Reimers, F., *The Basic Principles for Pharmacopoeial Tests*, Heinemann Medical Books, London, 1956, p. 50.
2. Zimmermann, J., Krogh-Svendsen, E., and Reimers, F., Limit tests for impurities IV. Investigations into the reproducibility of precipitation part III, *Analytica Chimica Acta*, 16, 1957, 6.
3. Evers, N., The detection of small quantities of calcium, *Analyst*, 56, 1931, 293.
4. Brause, G., Der nachweis des kalziums in aluminiumsalzen nach dem D.A.B. 6, *Pharmazeutischer Zeitung*, 73, 1928, 454.

6.4   Chlorides (European Pharmacopoeia 2.4.4)

The test ensures that the substance to be examined has a total content of chloride (Cl⁻), which is below the limit defined in the individual monograph. The test is referenced in several hundred monographs and this makes it one of the most highly referenced general tests. Chloride, obviously, is nontoxic and chemically inert and therefore harmless both from a safety viewpoint and with respect to the chemical stability of the substance to be examined. So the reason for limiting its presence is, in the of majority of cases, that chloride is extensively present in raw material and reagents used in chemical manufacturing, and a chloride residue, therefore, is a sign of insufficient purification.

The limit test for chlorides is based on the same chemistry as the identification test for chlorides, 3.16. Chlorides. The opalescence given by precipitation of the chloride present in the substance to be examined with silver is compared to the opalescence given in a similar precipitation of a chloride standard of known concentration.

$$Ag^+ + Cl^- \rightarrow AgCl\downarrow$$

Silver chloride is an example of a soft crystal and is therefore not susceptible to the problems seen when precipitating hard crystals such as barium sulfate and calcium oxalate. This, however, does not mean that the opalescence obtained in a test or standard solution is independent of the operational parameters of the precipitation procedure. But it can be considered a more reproducible and rugged determination compared to hard crystal precipitations, and the steps in the procedure contributing to loss of reproducibility are more easily standardized. The most obvious difference is that a fairly reproducible test procedure can be obtained without the use of a seeded standard, as is the case in 6.3. Calcium and 6.12. Sulfates.

Fifteen ml of the prescribed solution is added to 1 ml of *dilute nitric acid R* and the mixture is cast down as a single addition into a test tube containing 1 ml of *silver nitrate solution R2*. A standard is prepared in the same manner using 10 ml of *chloride standard solution (5 ppm Cl) R* and 5 ml of *water R*. The tubes are examined laterally against a black background. After standing for 5 min protected from light, any opalescence in the test solution is not more intense than that in the standard.

The two major causes of poor reproducibility is the way in which the chloride containing test solution and the silver-containing reagent are brought together and the temperature of the solution the salt precipitates from. The time allowed from performing the precipitation to the evaluation of the result influences the result, and, to a lesser extent, the ionic strength of the test/standard solutions.

Several studies have shown the intensity of the opalescence obtained in a test solution with defined chloride, and silver content is affected by the way the chloride-bearing test solution and the silver-bearing reagent solution are mixed.[1-3] These investigations reveal that a method like the one dictated in the *European Pharmacopoeia*, where the reagents are poured together in one fast movement, gives a relatively weak opalescence compared to what would have been obtained if, for example, the silver reagent had been added drop-wise. Results obtained with these two different methods can easily afford results that differ by a factor of two.

Since there is no problem in obtaining the necessary sensitivity, one would prefer the method with the best reproducibility and ruggedness, and in this respect the pouring method has a clear advantage. First, it is easier to reproduce a procedure where two solution are poured together than to reproducibly adding one solution to the other in a drop-wise fashion and doing some sort of mixing afterward. Also the reproducibility of the pouring method is better because the density of the two solvents does not affect the results. In the drop method the reagent falls through the sample solution and does part of the mixing, and then their relative densities suddenly become a factor.

Still, even when using the pouring technique, it must be emphasized that the test gives a result sensitive toward the handling of the reagents, so care should the exercised. Also is it evident that comparing a sample to a standard prepared by another person is not a good practice. Even the qualitative appearance of the precipitate can be influenced by the manner of precipitation. A precipitate brought about by the pour method will often have a faintly blue tone, whereas a slow precipitation often gives a more gray appearance.

One reference has investigated thoroughly the influence of temperature and ionic strength on the opalescence obtained. As illustrated in Table 6.1, temperature has a marked influence in the pour method in the way that a higher temperature gives a higher nephelometric value.

The nature of the nephelometric value is described in Chapter 4, "Precipitation in Limit Tests." Contrary to these findings there is a more limited effect of performing the determination in test solution containing 5% of various salts.[4]

Table 6.1 Effect of Temperature
on Nephelometric Value

Temperature	6°C	17°C	32°C
Nephelometric value	130	164	196

Table 6.2 Effect of Various Salts on Nephelometric Value

Salt present	No salt	$ZnSO_4$	Na_2SO_4	KNO_3
Nephelometric value	137	156	169	170

When the precipitation is performed in a nitric acid environment one aim is to enhance the selectivity of the test as is discussed below, but besides this it will enhance the sensitivity of the test by giving a stronger opalescence at equivalent chloride concentrations.[5] Another positive effect is that a silver chloride precipitation from an acid solution is less sensitive toward variations in reagent amounts used and presence of farren salts, as compared to one from a neutral solution.[6] Still the presence of some ions can influence the precipitation and especially the presence of KNO_3, Na_2SO_4, and $ZnSO_4$ and will give a somewhat enhanced opalescence as shown in Table 6.2.

The selectivity of the test is quite limited, even compared to the specificity seen in the identification test for chlorides. In the identification three criteria have to be fulfilled to qualify for a positive reaction. The unknown should give a white (curdled) precipitate formed upon addition of silver nitrate, which is insoluble in dilute nitric acid but redissolves in ammonia. In the limit test 2.4.4. Chlorides any substance capable of giving a white or weakly colored precipitate in dilute nitric acid will give a response like chloride, and this should be remembered in case of an unexpected result. For the sake of example the following ions and substances are capable of giving a false positive reaction: bromide, iodide, bromate, iodate, sulfite, chlorate, oxalate, and benzoate. In addition to this a variety of more complex organic substances are likely to precipitate, for example, alkaloids.

Besides the risk of falsely ascribing a response in the test to a content of chloride, there is a substantial risk of interference caused by contamination by various substances in the test solutions, forming insoluble silver salts. Reagents concentrate in the standard (16 ml) and in the test solution at equal intensity.

References

1. Kleinman, H., Über die bestimmung der phosphorsäure IV. Die bestimmung der phosphorsäure als strychnin-phosphorsäure-molybdänverbindung (nephelometrie). 1. Allgemeine prinzipien der nephelometrie und konstruktion eines neuen nephelometers, *Biochemische Zeitschrift*, 99, 1919, 140.
2. Scott, A.F. and Moilliet, J.H., The preparation of nephelometric test solutions, *J. Amer. Chem. Soc.*, 54, 1932, 205.

3. Scott, A.F. and Hurley, F.H., A procedure for the preparation of uniform nephelometric suspensions, *J. Amer. Chem. Soc.*, 56, 1934, 333.
4. Reimers, F. and Gottlieb, K.R., *Contributions from the Danish Pharmacopoeia Commission*, Vol. 1, Grænseprøver for urenheder/Limit tests for impurities, Ejnar munksgaard, København, 1946, pp. 19, 172.
5. Kolthoff, I.M. and Yutzy, H., The nephelometric determination of chloride, *J. Amer. Chem. Soc.*, 55, 1933, 1915.
6. Thörn, N., Kvalitativ prövning för påvisande av klorid, *Farmacevtisk Revy*, 40, 1941, 38.

6.5   Fluorides (European Pharmacopoeia 2.4.5)

The test ensures that the substance to be examined has a total content of fluoride (F^-), which is below the limit defined in the individual monograph. Only three monographs reference fluorides; calcium hydrogen phosphate anhydrate, dihydrate, and fluorozapam hydrochloride. The two calcium salts are at risk of containing a fluoride residue from the raw material used in its production, in fluorozapam hydrochloride it can be regarded as an impurity of synthesis. The solubility of fluoride salts shows a somewhat untypical solubility profile since its alkaline earth salts are insoluble, whereas its silver, mercury, aluminum, and nickel salts are readily soluble. Those of lead, copper, iron, barium, and lithium are slightly soluble. Fluoride is the most electronegative of all elements and has a strong ability to form complexes especially with metals of a valence higher than one.

This strong complex-forming ability of fluoride has been widely utilized in a large number of colorimetric methods where the colored complex between a central metal ion and an organic ligand is faded when the fluoride added eliminates the ligand by displacement. The organic ligand was most often alizarin or a closely related derivative. The method used in the limit test of the pharmacopoeia, first described in 1959,[1] differs from these procedures in that the fluoride added participates in the metal chelate instead of deteriorating it. The procedure is still, however, performed with an alizarin derivative, aminomethylalizarindiacetic acid (Figure 6.5.1). To minimize interference the fluoride present in the substance to be examined is first separated from the substance to be examined itself, and from any potential interfering ions present in the test solution. This is done in a distillation step

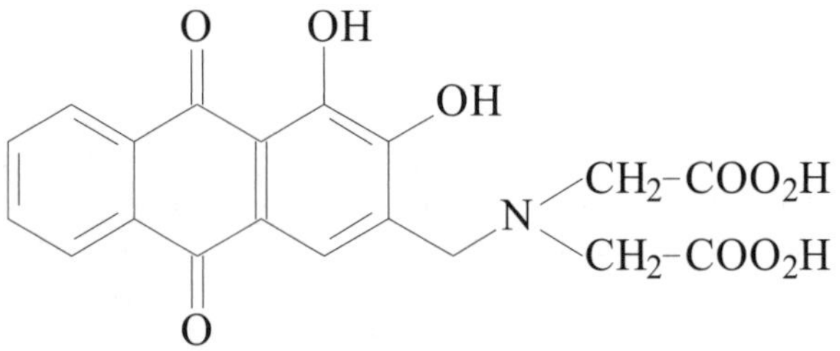

Figure 6.5.1 Aminomethylalizarindiacetic acid.

through volatilization of fluoride following the same mechanism as used in the identification test 3.29. Silicates, a procedure first described by Willard and Winter in 1933.[2]

The test is performed using a glass apparatus described precisely in the monograph, which is commercially available from several suppliers.

To the inner tube of the apparatus the prescribed quantity of the substance to be examined, 0.1 g of acid-washed *sand R* and 20 ml of a mixture of equal volumes of *sulfuric acid R* and *water R* is introduced. The jacket containing *tetrachloroethane R* is maintained at its boiling point (146°C). The steam generator is heated and a distillation is performed, collecting the distillate in a 100 ml volumetric flask containing 0.3 ml of *0.1 M sodium hydroxide* and 0.1 ml of *phenolphthalein solution R*. A constant volume (20 ml) in the tube is maintained during distillation and it is ensured that the distillate remains alkaline by adding *1 M sodium hydroxide* if necessary. The distillate is diluted to 100 ml with *water R* (test solution). A standard is prepared in the same manner by distillation, using 5 ml of *fluoride standard solution (10 ppm F) R* instead of the substance to be examined.

The fluoride present in the inner tube will, in the acidic environment, react with the silicium dioxide of the quartz sand forming hexafluorosilicic acid.

$$SiO_2 + 6H^+ + 6F^- \rightarrow H_2[SiF_6] + 2H_2O$$

But the dehydrating sulfuric acid present and the heat applied will instantly change it into volatile silicium fluoride.

$$H_2[SiF_6] + heat - H_2O \rightarrow SiF_4\uparrow$$

The silicium fluoride gas is led with the steam current and condenses in the 100 ml volumetric flask. Here, as a result of the reaction below, fluoride is again converted into hexafluorosilicic acid.

$$3SiF_4\uparrow + 2H_2O \rightarrow SiO_2\downarrow + 2[SiF_6]^{2-} + 4H^+$$

But, in the alkaline conditions of the distillate collector, hexafluorosilicic acid liberates its fluoride content leaving the silicium as insoluble silicic acid.

$$[SiF_6]^{2-} + 4OH^- \rightarrow H_2SiO_3\downarrow + 6F^- + H_2O$$

Any fluoride present in the substance to be examined is thereby now separated from interfering and disturbing ions. However, if an excess of boron, aluminum, zirconium ions, or gelatinous silica is present in the inner tube, it might retain the fluoride ions.[3] The use of tetrachloroethane is intended to maintain the sulfuric acid sample solution at a temperature that gives a good volatilization of fluoride and does not allows too much of the sulfuric to distill over into the collector. The fluoride is now trapped in the alkaline collector solution, and in the second step of analysis the chelatization with aminomethylalizarindiacetic acid is performed.

Into two glass-stoppered cylinders, 20 ml of the test solution and 20 ml of the standard and 5 ml of *aminomethylalizarindiacetic acid R* are introduced. After 20 min, any blue color in the test solution (originally red) should not be more intense than that in the standard.

In the absence of cations, the reagent aminomethylalizarindiacetic acid is a pH indicator (Figure 6.5.2) which changes from yellow to red at about pH 4.3 and to blue above 13.

The pH of the test solution offers that a yellow solution is expected but in the presence of cerium(III) a red complex of the structure (Figure 6.5.3) is formed. Most likely it has a red color because the electronic structure of the chelate is similar to the red form of the pure substance in pH up to about 10. When the fluoride is added a new structure (6.5.4) is formed. This is blue,

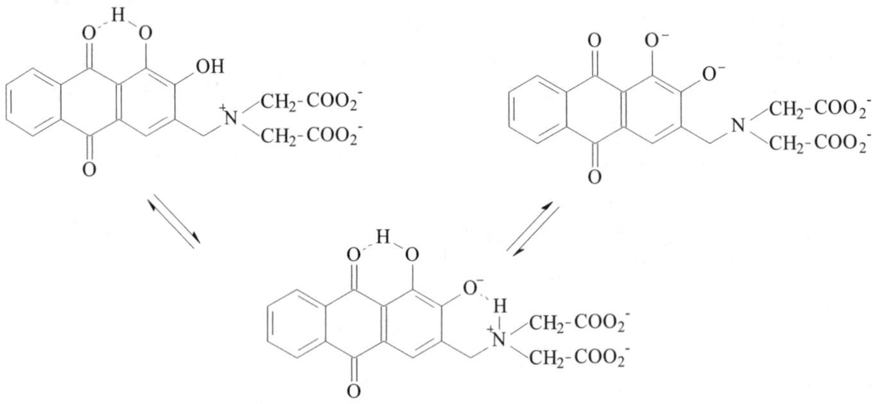

Figure 6.5.2 Aminomethylalizarindiacetic acid at different pH.

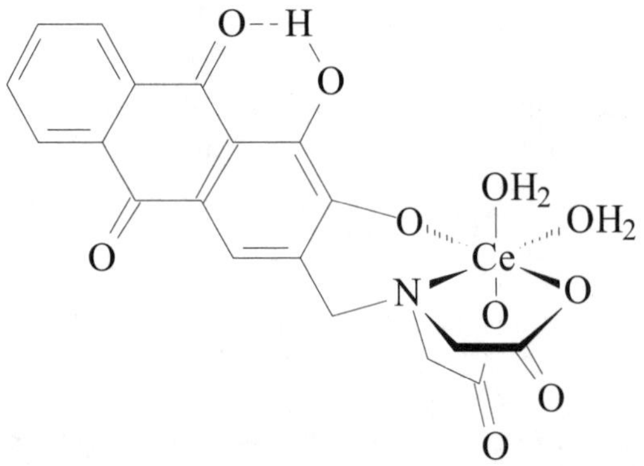

Figure 6.5.3 Cerium(III) complex.

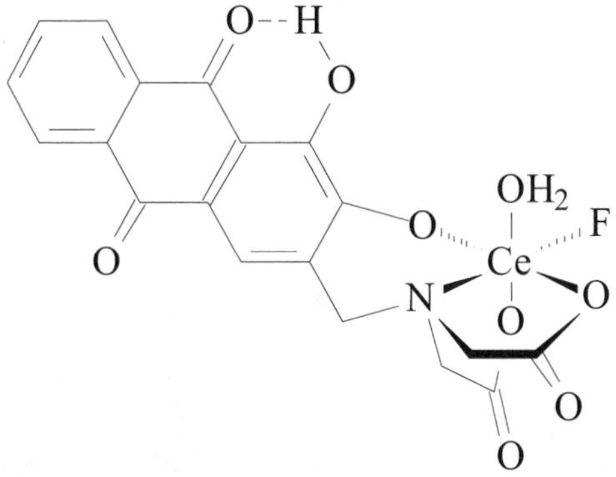

Figure 6.5.4 Cerium(III) fluoride complex.

most likely because its electronic structure is similar to the blue form of the pure substance.

Several other cations, such as calcium, barium, cadnium, nickel, magnesium, and mangane(II), are capable of forming a complex analogue to (Figure 6.5.3) thereby changing the color from yellow to red. But only the complex of cerium(III) will form by the addition of fluoride.

In quantitative tests based on this methodology, sulfate, which was a major cause of interference in many of the older spectrofotometric methods of fluoride determination, does not interfere and phosphate does likewise unless present in sixfold excess.

References

1. Belcher, R., Leonard, M.A. and West, T.S., Submicro-methods for the analysis of organic compounds. Part X. Determination of fluorine, *J. Chem. Soc.*, 1959, 3577.
2. Kolthoff, I.M. and Elving, P.J., *Treatise on Analytical Chemistry*, Part II, Vol. 7, Interscience Publishers, New York, 1961, p. 207.
3. Leonard, M.A. and West, T.S., Chelating reactions of 1,2-dihydroxyanthraquinon-3-ylmethyl-amine-NN-diacetic acid with metal cations in aqueous media. *J. Chem. Soc.*, 1960, 4477.

6.6   *Magnesium (European Pharmacopoeia 2.4.6)*

The test ensures that the substance to be examined has a total content of magnesium (Mg^{2+}), which is below the limit defined in the individual monograph. Magnesium is exclusively found as the bivalent cation Mg^{2+}. Its oxide, hydroxide, carbonate, and phosphate salts are insoluble in water, whereas the rest of its common inorganic salts are water soluble. Because of its insoluble hydroxide and carbonate salts, magnesium has a water solubility that is closely linked to solution pH.

The test is based on the precipitation of the inner complex that magnesium ion forms with the organic reagent hydroxyquinoline (Figure 6.6.1), a reagent that is also known as 8-hydroxyquinoline or oxine. The use of this reagent was first described in 1881 by Z.H. Skraup, but the pharmacopoeia uses a special version of the procedure where selectivity is enhanced by extracting the precipitate into chloroform with the aid of butylamine.

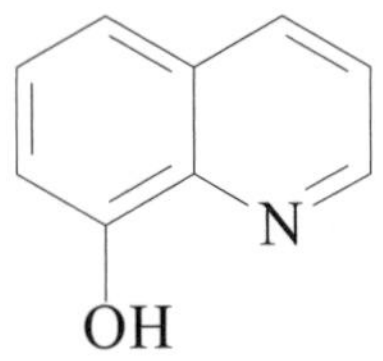

Figure 6.6.1 Hydroxyquinoline.

The selectivity of hydroxyquinoline in itself is very poor. One reference, which deals with the reagent in general and also its use for magnesium determinations in some detail, has a table (Table 6.3) of cations that can be precipitated by the reagent at different pH values.[1] The lack of selectivity is also underlined in another reference, which states that the reagent can be used for identification of magnesium, but that all other metals except sodium and potassium must be absent.[2]

In the limit test of the *European Pharmacopoeia*, butylamine (Figure 6.6.2) is added since it reacts with the precipitate to form a complex that, contrary to the pure magnesium hydroxyquinoline precipitate, is soluble in apolar solvents like chloroform.

In the first step of analysis 10 ml of the prescribed solution is added to 0.1 g of *disodium tetraborate R* to buffer the solution. Adjust the solution, if necessary, to pH 8.8 to pH 9.2 using dilute *hydrochloric acid R* or dilute *sodium hydroxide solution R*. Shake with two quantities each of 5 ml, of a 1 g/l solution of *hydroxyquinoline R* in *chloroform R*, for 1 min each time. Allow to stand. Separate and discard the organic layer.

This preextraction step eliminates a high number of cations forming extractable complexes with hydroxyquinoline in the pH interval given by the tetraborate buffer. The list of cations includes copper, calcium, gallium, aluminum, lead, bismuth, thorium, and to some extent zinc and cadium. Other cations including nickel and iron, which is extracted as well, but this

Table 6.3 Cations Precipitated by Hydroxyquinoline at Different pH.

pH interval	Cation
2–6	Ag, Al, Cd, Co, Cu, Fe, Ga, Hf, Hg, In, Mo, Nb, Ni, Pd, Ta, Th, Ti, U, W, Zn, Zr
6–10	Ac, Al, Be, Cd, Cu, Fe, Ga, Hf, Hg, In, La, Mg, Mn, Nb, Sc, Ta, Th, Ti, U, Y, Zn, Zr
>10	Cd, Cu, Fe, Mg, Zn

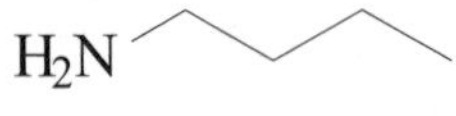

Figure 6.6.2 Butylamine.

Table 6.4 Interferences at Different pH

Interfering compound	% mg extracted		
	pH 10.5–11.0	pH 11.0–11.5	pH 12–13
NH_4F	84–71	71	59–45
Sodium acetate	>56	100	68
Sodium oxalate	100	100	72
Sulfo salicylic acid	~0	15	19–3
Potassium cyanide	70–100	100	70–50
EDTA	0	0	0
Na_2SO_4	n.b.	100	n.b.
Na_2PO_4	75–100	100	n.b.
Sodium citrate	n.b.	100	n.b.

is less important since their presence would be revealed as they form complexes of other colors than the green-yellow obtained by the majority of cations.[3]

To the aqueous solution 0.4 ml of *butylamine R* and 0.1 ml of *triethanol-amine R* are added. The solution is adjusted, if necessary, to pH 10.5 to pH 11.5. Four ml of the solution of hydroxyquinoline in chloroform is added, shake for 1 min, and allow to stand and separate. The lower layer is used for comparison. A standard is prepared in the same manner using a mixture of 1 ml of *magnesium standard solution (10 ppm Mg) R* and 9 ml of *water R*. Any color in the solution obtained from the substance to be examined should not be more intense than that in the standard.

In the presence of butylamine, the magnesium hydroxyquinoline complex will form a new chloroform soluble complex, apparently of the composition $CH_3(CH_2)_3NH_3[Mg(Ox)_3]$, which gives the organic phase a yellow coloration. To further mask interfering cations triethanolamine is added to bind them in the water phase as ion pairs. The most important interfering ion masked by this procedure is aluminum.

The presence of a number of anions and complex binders will disturb the extraction and lower the amount of magnesium hydroxyquinoline complex reaching the chloroform phase as shown in Table 6.4.

The method is in most references described with a masking procedure using oxalate, cyanide, and H_2O_2.[1,2] The same reagent is used in the limit test for aluminum, but this test is not covered by the present book since it is a fluorescence method.

References

1. Fries, J. and Getrost, H., *Organische Reagenzien für die Spurenanalyse*, E. Merck, Darmstadt, 1975, pp. 22, 225.
2. Umland, F., and Hoffmann, W., Über die verteilung von metall-8-oxychino-lin-verbindungen zwichen wasser und organische lösungsmitteln. 3.Mitt. Photometrische bestimmung des magnesiums nach extraktion seines oxinates mit chloroform bei gegenwart von aminen, *Anal. chim. acta*, 17, 234, 1957.
3. Vogel, A.I., *Vogel's Textbook of Quantitative Chemical Analysis*, 5th ed., Longman Scientific & Technical, Essex, 1989, p. 140.

6.7 Magnesium and alkaline-earth metals (European Pharmacopoeia 2.4.7)

The test ensures that the substance to be examined has a total content of alkaline-earth metals, which is below the limit defined in the individual monograph. Originally, calcium, strontium, and barium were called alkaline-earth metals because of their ability to form basic oxides. Earth is an old word for oxides. Nowadays, the term is most often used for the Group II metals in the periodic table. The selectivity of the limit test suggests that the original definition be applied in the present monograph.

Table 6.5 The Group II Elements

H																	He
Li	Be											B	C	N	O	F	Ne
Na	Mg											Al	Si	P	S	Cl	Ar
K	Ca	Sc	Ti	V	Cr	Mn	Fe	Co	Ni	Cu	Zn	Ga	Ge	As	Se	Br	Kr
Rb	Sr	Y	Zr	Nb	Mo	Tc	Ru	Rh	Pd	Ag	Cd	In	Sn	Sb	Te	I	Xe
Cs	Ba	La	Hf	Ta	W	Re	Os	Ir	Pt	Au	Hg	Tl	Pb	Bi	Po	At	Rn
Fr	Ra	Ac	Unq	Unp	Unh	Uns	Uno	Unn									

The alkaline earth-metals are valence II cations, which form soluble chlorides and nitrates, whereas their carbonates, sulphates, phosphates, and oxalates are insoluble. At the present five monographs references magnesium and alkaline-earth metals, and all monographs describe simple halogens of potassium, sodium, and ammonium. The limit described varies between 100 ppm and 200 ppm calculated as calcium. Potassium, sodium, and the halogens have their origins from natural sources like seawater or through mining where alkali-earth metals are present.

The test makes use of a complexiometric titration with the chelator ethylenediamine tetra-acetic acid (EDTA) using Eriochrome Black as a metallochromic indicator.

EDTA (Figure 6.7.1) forms stable complexes with nearly all polyvalent metal cations and even some monovalent too. It is a multidentate complex binder capable of acting as six ligands on the same central ion (Figure 6.7.2).

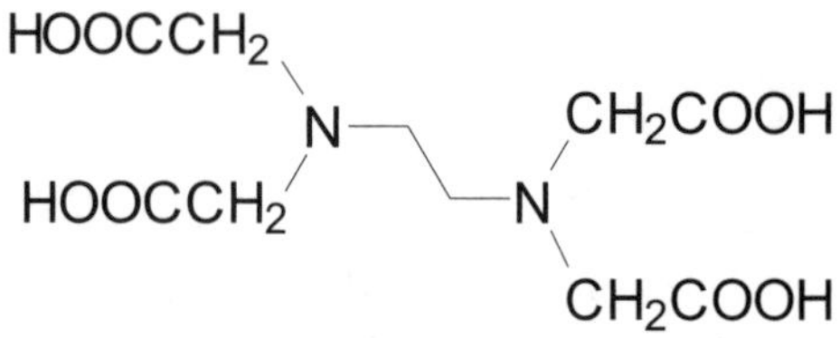

Figure 6.7.1 Ethylenediamine tetra-acetic acid (EDTA).

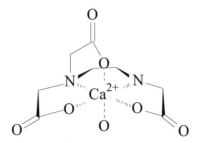

Figure 6.7.2 EDTA calcium complex.

The equilibrium of this complex formation and the stability constant of the complex are shown below.

$$M^{n+} + Y^{4-} \leftrightarrow MY^{n-}$$

$$K_{(absolute)} = [MY^{n-}]/[M^{n+}] \times [Y^{4-}]$$

This stability constant deals only with the species of EDTA where all carboxylic acid groups are deprotonated. The pKa values for these groups are pKa_1 2.0, pKa_2 2.67, pKa_3 6.16, and pKa_4 10.3, so the fully deprotonated EDTA dominates only in very alkaline solutions with pH above 12. In solutions of lower pH one has to calculate $[Y^{4-}]$ at the actual pH and use this concentration in calculating a stability constant. Similarly, if the metal M participates its other complexes, one has to calculate the $[M^{n+}]$ actually available for EDTA complexation. The stability constants incorporating the effects of pH and other complex reaction are called apparent stability constants. The absolute stability constant for a number of cations has been determined and is tabulated in Table 6.6.

Table 6.6 Absolute Stability Constants

Cation	log K	Cation	Log K
Ag^+	7.3	Cd^{2+}	16.46
Ba^{2+}	7.76	Zn^{2+}	16.50
Sr^{2+}	8.63	Pb^{2+}	18.04
Mg^{2+}	8.69	Ni^{2+}	18.62
Ca^{2+}	10.70	Cu^{2+}	18.80
Mn^{2+}	13.79	Hg^{2+}	21.80
Fe^{2+}	14.33	Cr^{3+}	23.0
Al^{3+}	16.13	Th^{4+}	23.2
Co^{2+}	16.31	Fe^{3+}	25.1

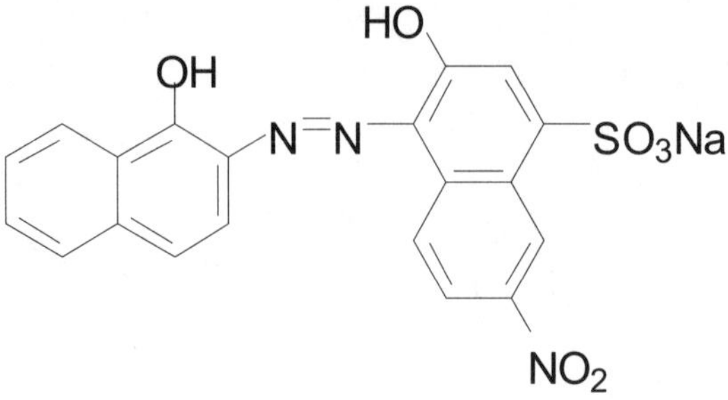

Figure 6.7.3 Eriochrome Black.

The indicator used in the test, *mordant black 11 triturate R*, is a dry dilution of Eriochrome Black (Figure 6.7.3), which is one of the most widely used metallochromic indicators. The acid form of the indicator is red, changes to blue at about pH 6, and to orange at pH 12. Only the blue is used as a metallochrome indicator, and it changes to red in the presence of metals.

It is very sensitive and gives a faint color under proper conditions with down to between 10^{-7} M and 10^{-8} M magnesium. It is not very stable in solution or a neat powder, and this is the reason for storing it dry diluted in sodium chloride. It gives a color change with the ions of magnesium, manganese, zinc, cadmium, mercury, lead, copper, aluminum, iron, titanium, cobalt, nickel, and platinum. The different metals and their different valances form complexes with the indicator of varying strengths. The complexation between the metal and Eriochrome Black can be explained quantitatively in exactly the same way as the metal EDTA complex.

$$M + I \leftrightarrow MI$$

$$K = [MI]/[M] \times [I]$$

M is the metal ion, I is the indicator, and MI is the metal indicator complex. All are shown without charge. The stability constant of the equilibrium, according to the law of mass action, has been determined for a number of cations (Table 6.7).

The complexes formed with Cu, Al, Fe(III), Ti(IV), Co, Ni, and Pt are so strong that EDTA are not able to release them from the indicator when added and the indicator is therefore blocked.

Table 6.7 Eriochrome Stability Constants

Cation	log K
Ba^{2+}	3.0
Ca^{2+}	5.4
Mg^{2+}	7.0
Mg^{2+}	12.9

In the first step of analysis 200 ml of *water R* is added 0.1 g of *hydroxyl-amine hydrochloride R*, 10 ml of *ammonium chloride buffer pH 10.0 R*, 1 ml of 0.01 M zinc sulfate, and about 15 mg of *mordant black 11 triturate R*.

The addition of *hydroxylamine hydrochloride R* aims to mask some of the potentially interfering ions by reducing to an oxidation state not capable of complexing the indicator (iron + copper + Mn). The buffer serves two purposes. First, it keeps pH at 10, which is the pH preferred by the indicator, throughout the titration. Second, being ammoniacal it complexes the zinc, thereby preventing it from precipitating as zinc hydroxide and also lowers zinc's apparent stability complex. This plays a vital role in the test as explained below.

$$Zn^{2+} + 4NH_3 \rightarrow [Zn(NH_3)_4]^{2-}$$

Then without having added any sample, heat to about 40°C and titrate with *0.01 M sodium edetate* until the violet color changes to full blue. The test solution prepared in step one contains both zinc and the indicator Eriochrome Black and it will form a violet complex. But when the test solution is titrated with EDTA in the second step of analysis the stronger EDTA complex will deplete zinc ions of the solution and the indicator complex making it change the color to its metal free blue form, since titration is carried out to blue and EDTA is complexed with zinc.

To this solution the prescribed quantity of the substance to be examined, dissolved in 100 ml of *water R* (or the prescribed solution), is added to the test solution. If the color of the solution changes to violet, it is titrated with *0.01 M sodium edetate* until the full blue color is again obtained. The volume of *0.01 M sodium edetate* used in the second titration does not exceed the prescribed quantity.

If the sample solution contains any ions of magnesium and alkaline-earth metals, they will replace zinc and form EDTA complexes since their apparent stability constants are much larger than zinc. The zinc set free will form a complex with the indicator changing the color to violet. The amount of zinc replaced is equivalent to the amount of magnesium, and alkaline-earth metals in the sample solution is determined in the second titration.

At the surface, one should not expect magnesium and calcium to be able to replace zinc in its EDTA complex since the absolute stability constant of zinc is several orders of magnitude higher than the absolute stability constant of the Group II metals. This can be explained by the presence of ammonium ions and the relevant ion stability constants with Eriochrome Black.

First, zinc will form a relatively stable complex with ammonium and this means that not all the zinc added is available for EDTA complexation, giving an apparent stability constant considerably lower than the absolute stability constant. Second, zinc forms a very stable complex with Eriochrome Black and the indicator will therefore be able to pull zinc out of its EDTA complex even if only a weak competitor for EDTA as the calcium ion is introduced. The reason for adding zinc for the purpose of replacement is that out of the ions to be examined only magnesium has a high enough stability constant with the indicator to give a color response by itself. The others will not form a colored complex with Eriochrome Black, but rely on their ability to replace zinc.

Selectivity in an EDTA titration is given by the selectivity of EDTA and the selectivity of the indicator. EDTA is as mentioned very unspecific, and not many cations can be ruled out except some monovalent ions including most notably the alkali metals. The selectivity of the indicator gives a list of possible interfering cations including manganese, cadmium, mercury, lead, copper, aluminum, iron, titanium, cobalt, nickel, and platinum. Manganese, iron, copper, aluminum, titanium (IV), cobalt, nickel, and platinum block the indicator and would reveal themselves by not allowing a color change back to blue by adding EDTA. Cadmium, mercury, and lead act very much as zinc and magnesium and will behave like those in the titration. Their presence will, however, be revealed by the limit test in 6.8. Heavy metals. All monographs currently referencing magnesium and alkaline-earth metals also reference 6.8. Heavy metals.[1–3]

References

1. Schwarzenbach, G. and Flaschka, H., *Die komplexometrische Titration*, 5. Auflage, Ferdinan Enke Verlag, Stuttgart, 1965, several chapters.
2. Flaschka, H.A., *EDTA Titrations*, 2nd Ed., Paragamon Press, Oxford, 1964, several chapters.
3. Vogel, A.I., *Vogel's Textbook of Quantitative Chemical Analysis*, 5th ed., Longman Scientific & Technical, Essex, 1989, several chapters.

6.8   Heavy metals (European Pharmacopoeia 2.4.8)

The test ensures that the substance to be examined has a content of heavy metals, which is below the limit defined in the individual monograph. Heavy metals are in the test defined by the selectivity of the methods to include lead, copper, silver, mercury, cadmium, bismuth, ruthenium, gold, platinum, palladium, vanadium, arsenic, antimony, tin, and molybdenum.

The contents of heavy metals in active pharmaceutical ingredients and excipients are tested for two reasons: due to the toxicity of the elements, but also because a heavy metal residue is a general marker for the quality of the product. The toxicologically acceptable content of heavy metals in a given substance to be examined is set taking into account its route of administration, size of daily intake, and typical duration of intake. Such calculations impart that a substances taken orally for a short duration in a very low dose could safely be allowed a heavy metal content above 1,000 ppm, whereas a substance given in high concentration through parenteral administration for a long duration should have a limit below 1 ppm.

Many active pharmaceutical ingredients fall in the former category where the toxicologically defined limit becomes meaninglessly high seen from a quality point of view. Likewise, many excipients fall in the latter category where the limit becomes meaningless from an analytical point of view since the limit would be far below the methods limit of detection. In both cases it is often still considered relevant to have a limit of heavy metal content, but then more as a marker of the general product quality.[1]

This limit test for heavy metals is one of the most widely used and is being referenced in many monographs. The methods most often referenced are method (a) and method (c).

Active pharmaceutical ingredients and excipients can in principle be contaminated by heavy metals through various routes, for example, via the extraction of the substance to be examined from contaminated herbal material or by use of contaminated raw material of mineral origin, through handling in metallic chemical synthesis and pharmaceutical production equipment, and by use of metal containing contaminated reagents in chemical synthesis. The testing of heavy metal residuals in herbal drugs and fatty oils are excluded for the purpose of this discussion since they are controlled using another method. Several monographs of the *European Pharmacopoeia* include specific tests for individual heavy metal cations besides referencing heavy metals 2.4.8.

The *European Pharmacopoeia* was born with a heavy metal test inherited from the national European pharmacopoeias and the methodology applied has been known for centuries. One of the world's oldest pharmacopoeias, *The Unites States Pharmacopoeia*, included a general test for heavy metals in volume VIII from 1905. The aim of the test was, through a sulfide precipitation in a both strongly acidic and alkaline medium, to detect the presence of undesirable metallic impurities. These were antimony, arsenic, cadmium, copper, iron, lead, and zinc. By that time heavy metal salts were widely used in therapy, and the aim of the test was therefore to a higher extent to reveal mislabeled products more than to reveal heavy metal contamination. This was changed in the 1942 *USP* XII, where a lead containing standard was included in the test and precipitation was performed only under mildly acidic conditions. The aim was now to detect poisonous heavy metal residuals, most notably lead and copper, since these metals were widely used in production equipment at the time, making this testing strategy rational. Other metal often used nowadays, on the contrary, like iron, chromium, and nickel, are unfortunately not revealed by the test.[2]

The chemical basis for the determination is the formation of a colloidal precipitate of insoluble heavy metal sulfide salt, as shown below using lead as an example.

$$Pb^{2+} + H_2S \rightarrow PbS\downarrow + 2H^+$$

Originally these determinations were often carried out by bobbling hydrogen sulfide gas through the sample preparation, giving a saturated solution containing approximately 0.4% in neutral water. The concentration of saturation is lower in acidic solutions and higher in alkaline solutions due to the

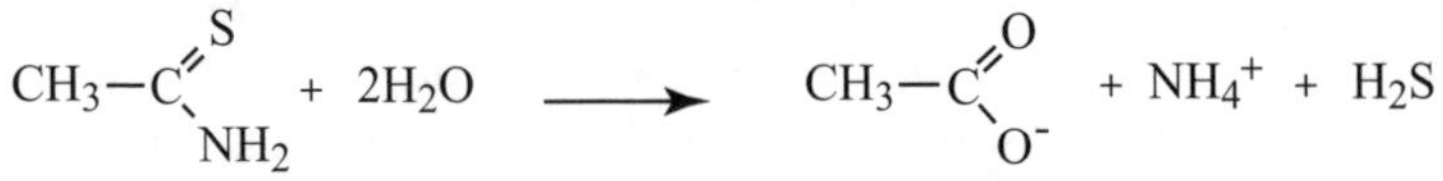

Figure 6.8.1 Thioacetamide hydrolysis.

formation of nonwater-soluble ionic sulfide, S^{2-}. Hydrogen sulfide is a diprotic acid capable of releasing two protons.

$$H_2S \rightarrow HS^- + H^+ \rightarrow S^{2-} + 2H^+$$

The sulfide ion forms highly water insoluble salts with many cations. The fact that the sulfide ion concentration is pH dependent has often been used for separating various cations through selective precipitation and for identification of unknown cations by group separation. Some cations form sulfide salts so insoluble that they precipitate in a hydrogen sulfide saturated solution even at a pH below 1 where the sulfide solution is very low. These are in group separation normally referred to as the sulfide group subdivided into the copper and arsenic groups. The order of precipitation as a function of pH in the low pH group is: arsenic, molybdenum, silver, copper, antimony, bismuth, mercury, gold, platinum, tin, cadmium, lead. Some less insoluble sulfide salts will not precipitate before neutrality or even under alkaline conditions. About one-third of all cations can form insoluble sulfides.

Since hydrogen sulfide is poisonous, it has been substituted with the reagent thioacetamide. This reagent is stable in the neutral conditions of the *thioacetamide solution*, but will hydrolyse when mixed with the *IM sodium hydroxide* to yield a solution saturated with hydrogen sulfide.

To ensure a quantitative hydrolysis the prescribed boiling is necessary, but if the boiling is prolonged, elemental sulfide might precipitate. The reagent contains glycerine in order to enhance the observation by slowing down hydrolysis, but also by stabilizing the colloidal suspension in the test solution.[3]

Methods (a), (b), (c), and (d) are identical except for the sample preparation used. Method (e) has been developed for substances where a limit below 5 ppm heavy metal is dictated, and method (f) is a wet digesting procedure developed to replace the somewhat tedious dry ashing procedures of methods (c) and (d).

Method (a)

To 12 ml of the prescribed aqueous solution 2 ml of *buffer solution pH 3.5 R* is added. The blend is mixed and added to 1.2 ml of *thioacetamide reagent R*. Mix immediately. A standard is prepared in the same manner using a mixture of 10 ml of *lead standard solution (1 ppm or 2 ppm Pb) R*, as prescribed, and 2 ml of the solution to be examined. A blank is prepared, using a mixture of

10 ml of *water R* and 2 ml of the solution to be examined. Compared to the blank, the standard shows a slight brown color. After 2 min, any brown color in the test solution should not be more intense than that in the standard.

These conditions will, as discussed above, ensure that the lead content of the standard and the content of any of the heavy metals mentioned in the first paragraph in the substance to be examined precipitates as sulfide salt. The pH given by the buffer defines the selectivity of the test as described above, but a discrepancy of the pH in between the standard solution and test solution can also give a difference in the intensity and color of the precipitate formed. Further deviations are expected when comparing different metal sulfides since they have somewhat different colors. Lead sulfide together with the sulfides of ruthenium, gold, platinum, molybdenum, and palladium are brown, whereas the sulfides of silver, mercury, bismuth, and copper appear black. Sulfides of cadmium, arsenic, antimony, and tin are yellow. The reason for testing in a low analyt concentration in the is that these color differences are then not very marked, whereas in higher concentration it would not be possible to compare all these cations with lead. But the color difference of the metal sulfides claimed to be included in the scope of the test has often been a point of criticism for users of the test.[4]

The appearance of the precipitate in respect to both color and intensity is also affected by the ionic strength of solutions, a fact that is taken into account in the test by adding sample solution to the standard and blind solutions. However, because cationic sulfides' sensitivity in appearance toward pH are different, they do not behave alike when their ionic strengths differ. They do not behave alike on differences in ionic strength. The most common picture, however, is the higher ionic strength gives more coloration. Some of the heavy metals form a sulfide precipitate where the speed of reagent mixing gives a difference in the size of the particle and thereby the appearance of the colloid suspension.

The reason for including a blind solution in the test is to ensure that the water used is of a quality that does not cause interference in the test but also to make it possible to assess and ensure that the standard is colored.

Method (b)

This procedure is identical to method (a) except that the substance to be examined and the *lead standard solution (100 ppm Pb)* is diluted in an organic solvent containing a minimum percentage of water. An example given in the text is dioxan containing 15 percent of water or acetone containing 15 percent of water. This method is intended for substances that are not sufficiently water soluble to afford colorless solution with water.

The key element in a judgment of method (b)'s suitability to a given substance is the ability of the test to give a reproducible sulfide precipitation in the solvent capable of dissolving the substance to be examined. Obviously it must also be able to dissolve a given metal content. Another prerequisite is that the substance to be examined in the given solvent neither interferes with the precipitation nor forms a chelate with the heavy metals present, making them unavailable for the precipitation.

Method (c)

Method (c) was originally made as an alternative using magnesium sulfate in the incineration to method (d) using magnesium oxide. The need for this was that it was earlier difficult to obtain a magnesium oxide of sufficient quality with regard to heavy metal content.[5] Only issues relevant for method (c) are dealt with here and the rest are discussed in method (d).

The prescribed quantity (not more than 2 g) of the substance to be examined is placed in a silica crucible with 4 ml of a 250 g/l solution of *magnesium sulfate R* in dilute *sulfuric acid R*. The sample is mixed using a fine glass rod. Heat cautiously. If the mixture is liquid, this is evaporated gently to dryness on a water-bath. Progressively heat to ignition and continue heating until an almost white or at most grayish residue is obtained. The ignition is carried out at a temperature not exceeding 800°C. Allow to cool. The residue is moistened with a few drops of dilute *sulfuric acid R*. Evaporate, ignite again, and allow to cool. The total period of ignition must not exceed 2 h. The residue is taken up in two quantities, each of 5 ml, of dilute *hydrochloric acid R*. Add 0.1 ml of *phenolphthalein solution R*, then concentrated *ammonia R* until a pink color is obtained. The sample is cooled, *glacial acetic acid R* is added until the solution is decolorized, and then 0.5 ml is added in excess. Filter if necessary and wash the filter. Dilute to 20 ml with *water R*. Of these 20 ml, 12 ml is taken as the prescribed solution upon which the procedure of method (a) is performed.

In the first step of the analysis, the substance to be examined is mixed with solutions of magnesium sulfate and sulfuric acid, and the solvent (water) is then removed by evaporation. This should be carried out with some patience to avoid sample or standard loss from spurts of the sample. Patience is also necessary in the next step of analysis when the heat progressively increased until ignition.

Some substances do not char unless they are intimately in contact with the magnesium sulfate and sulfuric acid and for this reason dilute sulfuric acid is added again.

Method (d)

In this method, a sample and test solution like the one of method (a) is obtained by incinerating the substance to be examined, which in this case will be an organic molecule. This sample preparation principle has some inborn problems of which the biggest are problems in achieving a complete incineration and evaporation of the organic sample molecule since this is a prerequisite for obtaining a clear sample solution. This subject is also dealt with in some detail in the 6.13. Sulfated ash.

In a silica crucible, the prescribed quantity of the substance to be examined is mixed thoroughly with 0.5 g of *magnesium oxide R1*. Ignite to dull redness until a homogeneous white or grayish-white mass is obtained. If after 30 min of ignition the mixture remains colored, allow to cool, mix using a fine glass rod and repeat the ignition. If necessary repeat the operation. Heat at 800°C for about 1 h. The residue is taken up in two quantities, each of 5 ml, of a mixture of equal volumes of *hydrochloric acid R1* and *water R*. Then add 0.1 ml of *phenolphthalein solution R* and concentrated *ammonia R* until a pink color is obtained. After the solution has cooled, *glacial acetic acid R* is added until the solution is decolorized and then 0.5 ml is added in excess. Filter if necessary and wash the filter. Dilute to 20 ml with *water R*.

As is the case with other ashing procedures, there are a number of pitfalls that have mostly to do with practical details in the procedure. Different organic substances behave very differently in the incineration process. Some melt and others char into a compact mass and some give off fumes that can easily catch fire. Especially when using a Bunsen burner, great care should be exercised to avoid spurts and fires since this obviously leads to a loss of analyt. When an almost white or at the most grayish residue is obtained the carbon is evaporated. But this evaporation is often not possible if degradation of the substance to be examined has not been completed all the way down to the point of a gray/white ash but only into a state of a black char.

The total period of ignition at 800°C must not exceed 2 h since such prolonged heating will lead to a loss in heavy metal. Even when this rule is followed a relatively high amount of some of the lower boiling heavy metals are lost, giving recoveries between 0% and 70%.[6] This is to some extent compensated because the standard is treated like a sample when it ideally should suffer a proportional loss. Also it is an explicit demand to the test that the standard must be darker than the blind, a procedure that will reveal an extensive loss of lead from the standard.

When evaporation is successfully completed the heavy metal residue is taken up by hydrochloric acid in which it easily dissolves. It is however often seen that the heavy metal is trapped in the bottom of the crucible in a thin hard layer of miscellaneous residue material from the incineration from which it has to be liberated by scratching with a glass rod.

Method (e)

The intent of method (e) is to provide a test with a limit of detection below what is achievable with method (a) to (d) and method (f). The method is claimed to have a limit of detection below 1 ppm. This is achieved by preparing a standard and sample of a higher heavy metal content then used in the other methods, prefiltering the standard and sample solution to a completely clear solution prior to the precipitation step, and finally collecting through filtration the precipitate formed as a concentrated spot on a filter disc.

Method (f)

The low recovery rate seen in methods (c) and (d) has led to the development of method (f), which is a wet digestion much in the line of the classical Kjeldal digestion. In this method good recovery rates are obtainable, but not without some limitations and drawbacks. The most obvious is that digestion of some substances is quite time consuming, but also evaporation of the digestion solvent can take some time.

It is clear from the many articles published in general scientific literature, in pharmacopoeia publications, and even in the *European Pharmacopoeia's* guide for the general test in the new monograph, that the tests of heavy metals method (a) to (f) have certain limitations. The most important are: the limit of detection is in general close to the limits of allowed content; comparing sulfide salts of different colors causes problems even though the amount of precipitate is very low; certain heavy metals are lost to a high extent during the incineration procedure of methods (c) and (d); the result obtained in methods (c) and (d) puts high demands on the skills of the operator carrying them out; and the wet ashing procedure of method (f) is for many substances very time consuming.

References

1. Council of Europe, *Pharmeuropa, Technical Guide*, Council of Europe, Strasbourg, 1990, p. 218.
2. Ciciarelly, R. et al., Determination of metal traces — a critical review of the pharmacopoeial heavy metal test, *Pharmacopeial Forum*, 21, 1995, 1638.
3. Tadeusz, L. and Ramotowski, S., Amid kwasu tiooctowego (akt) jako odczynnik do, wykrywania i oznaczania zanieczyszcze metalami cikimi w preparatach leczniczych w wietle wymaga farmakopealnych, *Acta Pol. Pharm* 14, 1957, 185.
4. Gernez, G. and Moine, I., Réflexions sur l'essai des métaux lourds, *Pharmeuropa*, 1, 1989, 249.
5. Hartke, K., 2.4.8. Grenzprüfungen (10. Lfg. 1999), in *Arzneibuch-Kommentar zum europäische arzneibuch*, Band I *Allgemeiner teil*, Hartke, K. et al., Wissenschaftliche Verlagsgellschaft, Stuttgart/Govi-Verlag — Pharmazeutischer Verlag, Escbor.
6. Blake, B. K., Harmonization of the USP, EP, and JP heavy metals testing procedures, *Pharmacopeial Forum*, 21, 1995, 1632.

6.9 Iron (European Pharmacopoeia 2.4.9)

The test ensures that the substance to be examined has a total content of iron (Fe^{2+} or Fe^{3+}), which is below the limit defined in the individual monograph. An iron contamination could arise from the raw materials used in chemical manufacturing or from iron-containing process equipment. Since metal process equipment is widely used the iron in this section is a quite commonly applied test, referenced in more than 100 monographs. Iron is a typical transition metal and can therefore exist in more that one oxidation level. But out of these only iron(II) and iron(III) are stable and they are both quite common. Both ions form insoluble hydroxides (white and reddish-brown respectively) but iron(II) is very easily oxidized into iron(III) at neutral and alkaline conditions. Even dissolved oxygen affords this reaction, so solutions of iron(II) have to be acidic to be stable. A solution containing iron(II) gives a green appearance, whereas a solution of iron(III) is pale yellow.

The methodology of the test, which was first described in 1879 by R. Andreasch, is based on the colored product formed when iron(II) or iron(III) ions reacted with thioglycollic (Figure 6.9.1) acid in the presence of ammonia.

The prescribed quantity of the substance to be examined is dissolved in *water R* and diluted to 10 ml with the same solvent. Alternatively, 10 ml of the prescribed solution is used. Two ml of a 200 g/l solution of *citric acid R* and 0.1 ml of *thioglycollic acid R* are added. The solution is mixed, made alkaline with *ammonia R*, and diluted to 20 ml with *water R*. A standard is prepared in the same manner, using 10 ml of *iron standard solution (1 ppm Fe) R*. After 5 min, any pink color in the test solution is not more intense than that in the standard.

When iron(II) is mixed with thioglycollic acid in the presence of ammonia, a colorless or almost colorless complex of Figures 6.9.2 or 6.9.3 is formed with ammonia and thioglycollic acid in a 1:1 or 2:1 ratio. This complex is oxidized by the present oxygen to a pink complex of Figure 6.9.4. Apparently Figure 6.9.4 is formed by an oxidation of the central iron(II) to iron(III). This means that the reaction will not take place if it is carried out in degassed solvents and reagents.[1]

If iron is present in the substance to be examined in the form of iron (III), and this is by far the most likely, first it will be reduced to iron(II) by the thioglycollic acid through the reaction shown in Figure 6.9.5. The iron(II) formed will thereafter participate in the complexion as described above.

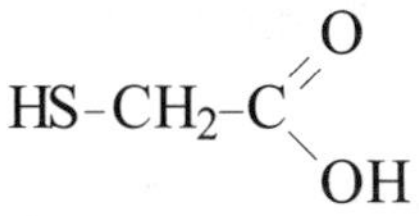

Figure 6.9.1 Thioglycolic acid.

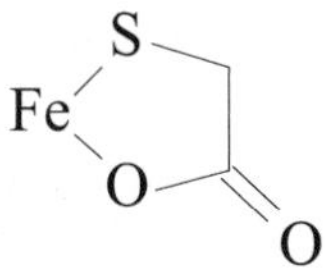

Figure 6.9.2 Colorless complex.

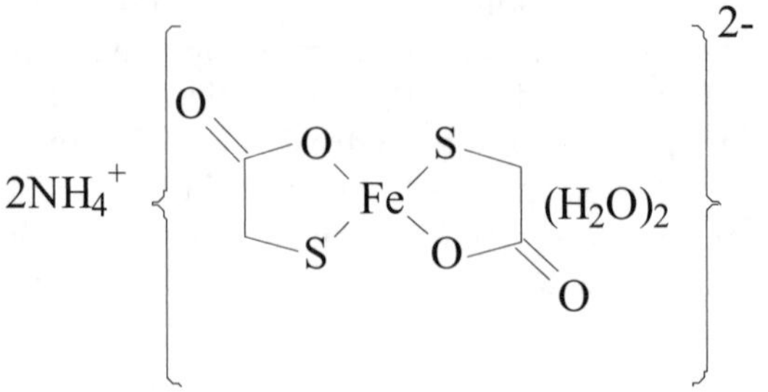

Figure 6.9.3 Colorless complex.

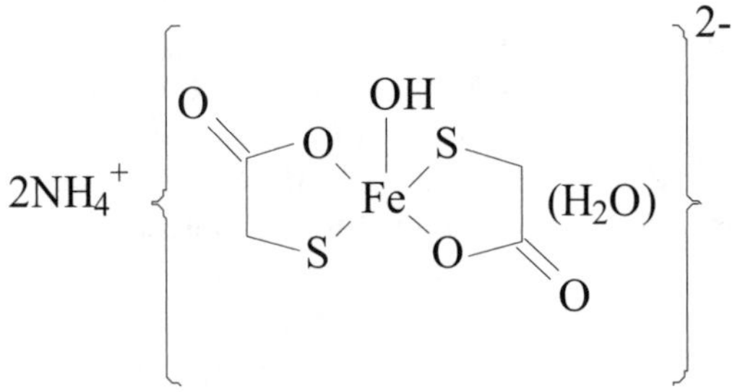

Figure 6.9.4 Pink complex.

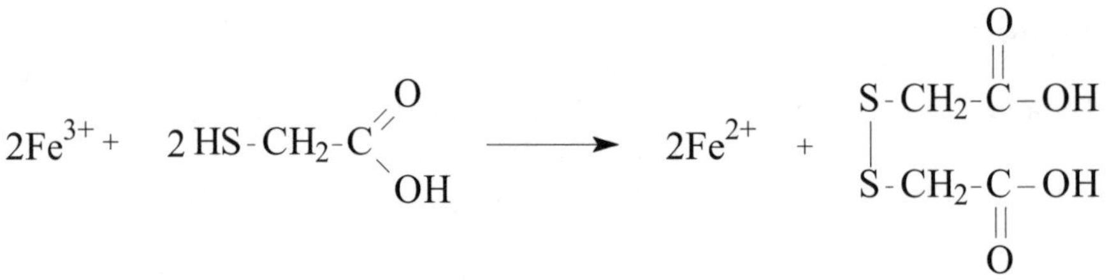

Figure 6.9.5 Iron(III) reduction.

The complexation is stated to take place in a pH interval of 6 to 11. At the pH of the test many cations could precipitate as oxides, hydroxide, or even carbonates (if present). The citric acid added prevents this through complexation of the cations. Some references state that the complex is sensitive to light but that a sufficient stability for up to 6 h is obtained if samples are kept in diffuse daylight.[2]

A number of ions are capable of giving a response, which to a varying degree will give a result similar to the one given by iron: silver, gold, bismuth, cobalt, mercury, lead, or uranium. If some ions are present together with iron, then the color obtained will be weaker: arsenic, cadmium, copper, tin, or zinc. Cyanide is said to disturb the determination. As in other cases these interferences are described in assays.[3,4]

References

1. Geffken, D. and Surborg, K.H., Eisen-thioglycolsure-reaktion. Theoretische und experimentelle betrachtungen, *Deutsche Apotheker Zeitung*, 128, 1235, 1988.
2. Lange, B. and Vejdělek, Z.J., *Photometrische analyse*, 1st ed., Verlag Chemie, Wienheim, 1980, p. 95.
3. Allport, N.L., *Colorimetric Analysis*. 1st ed., Chapman & Hall, London, 1945, p. 53.
4. Feigl, F., *Qualitative Analysis by Spot Tests*, 3rd ed., Elsevier Publishing Company, New York, 1947, p. 127.

6.10 Phosphates (European Pharmacopoeia 2.4.11)

The test ensures that the substance to be examined has a total content of phosphate (PO_4^{3-}), which is below the limit defined in the individual monograph. At the present less than 10 monographs reference phosphates, all of them describing chemical entities where phosphate has been used in the synthesis. An exception is the inclusion of the limit test in the sodium chloride monograph whose rationality apparently has been lost in time.[1] Of the remaining monographs a large group are the various salts of glycerophosphate (Figure 6.10.1) that is synthesized by adding phosphate to the carbon backbone. Phosphate can be found in a variety of oxidation levels and is capable of forming a large number of monomeric and polymeric oxo acids; therefore it should be specified that the methodology used in the present test is selective toward orthophosphates, PO_4^{3-}. A short introduction to some of the other phosphate compounds is given in 3.26. Phosphates (Orthophosphates).

The limit test is based on the specific coloration seen when ammonium phosphomolybdate is first formed and then reduced. Colorimetric methods for identifying and determining phosphates based on the formation of molybdate salts have been very popular and are intensively described in scientific literature. Many different modifications have been proposed, claiming advantages of different parameters in forming ammonium phosphomolybdate and utilizing different reductants, and none of the methods could be regarded as being unconditionally superior to the others. The method used in this section is different from the phosphate determination based on molybdate and vanadate used in the identification test in 3.26. Phosphates.

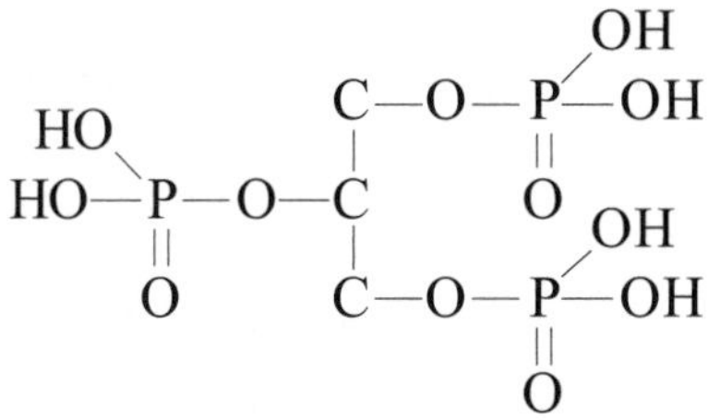

Figure 6.10.1 Glycerophosphate.

Lange and Vejdělek[2] reference more than 100 articles describing molybdate methods and 30 articles describing the methodology adapted in the European *Pharmacopoeia* were phosphate is first bound as ammonium phosphomolybdate and thereafter reduced by stannous chloride. These articles were published in the 1930s to the 1970s.

To 100 ml of the solution prepared (neutralized as prescribed if necessary), 4 ml of *sulphomolybdic reagent R3* is added. The solution is shaken and 0.1 ml of *stannous chloride solution R1* is added. A standard is prepared in the same manner using 2 ml of *phosphate standard solution (5 ppm PO₄) R* and 98 ml of *water R*. After 10 min, compare the colors using 20 ml of each solution. Any color in the test solution should not be more intense than that in the standard.

When under the conditions given the orthophosphate present in the sample solution reacts with the ammonium molybdate present in the sulphomolybdic reagent R3, yellow ammonium phosphomolybdate is formed in which Mo_3O_{10} replaces each of the oxygen atoms in phosphate.

$$HPO_4^{2-} + 3NH_4^+ + 12MoO_4^{2-} + 23H^+ \rightarrow (NH_4)_3[P(Mo_3O_{10})_4] + 12H_2O$$

Ammonium molybdate has the formula $(NH_4)_6Mo_7O_{24}$, but in the above reaction scheme, as well as in many other texts, it is presented as MoO_4^{2-} for simplicity. The product is in some cases presented with the Brutto formula, $(NH_4)_3[PMo_{12}O_{40}]$.

If the ammonium phosphomolybdate concentration is too high, a yellow precipitate is formed. This has often been used deliberately in test procedures, but such methods suffer from the reproducibility problems often encountered in tests based on precipitation. The environment must be acidic but the presence of nitric acid enhances the risk of ammonium phosphomolybdate precipitation, and hydrochloric acid gives a yellow precipitate of molybdic acid, H_2MoO_4. The presence of other ammonium salts also raises the risk of an unwanted ammonium phosphomolybdate precipitation provoked by the common ion effect.

In the second step, the reductant stannous chloride (tin(II) chloride) is added to the solution. The yellow ammonium phosphomolybdate is thereby reduced to a substance called molybdenum blue. The structure of this amorphous substance of ammonium, phosphate, and molybdate has not been elucidated, but the corresponding molybdate blue is presented with the formula $[(MoO_3)_{154}(H_2O)_{70}]^{y-}$.[3]

Many different reductants have been proposed. The good thing about tin(II) chloride is that it is fast compared to many of the less harsh reagents such as hydrazine sulphate. Thus it gives a rapid reaction already at room temperature.

Molybdate alone is not easy to reduce but when bound to phosphate it is. On the contrary, if the reducing power of the reductant added is too strong, molybdate itself can be oxidized to a blue product that after a while changes to green and finally brown.

Silicate (SiO_3^{2-}) and arsenate (AsO_4^{3-}) give a reaction similar to phosphate, thereby affording a false positive reaction.[4] A number of ions are reported to interfere when the test methodology is used in assays, but the relevance of these is believed to be limit in the limit test. The blue color formed is pH dependent.[5]

Much of the literature on the method deals with its use in assays, and for this purpose it is described as a good method. This is also shown in its widespread use and the many publications dealing with the method.

References

1. Hartke, K., Natrium chloride (10. Lfg. 1999), in *Arzneibuch-Kommentar zum europäische arzneibuch*, Band 3, Hartke, K. et al., Wissenschaftliche Verlagsgellschaft, Stuttgart/Govi-Verlag — Pharmazeutischer Verlag, Escbor.
2. Lange, B. and Vejdělek, Z.J., *Photometrische analyse*, 1st ed., Verlag Chemie, Wienheim, 1980, p. 375.
3. Müller, A. et al., Molybdänblau — ein 200 jahre altes geheimnis wird gelüftet, *Angewandte Chemie*. 1996, p. 1296.
4. Vogel, A.I. and Svehla, G., *Vogel's Qualitative Inorganic Analysis*, 6th ed., Longman Scientific & Technical, Essex, 1987, pp. 87 198, 201.
5. Kodama, K., *Methods of Quantitative Inorganic Analysis*, Interscience Publishers, New York, 1963, p. 430.

6.11   Potassium (European Pharmacopoeia 2.4.12)

The test ensures that the substance to be examined has a total content of potassium (K^+), which is below the limit defined in the individual monograph. The test is referenced in several hundred monographs, which makes it one of the most highly referenced general tests. Potassium, obviously, is nontoxic and chemically inert and therefore harmless both from a safety viewpoint and with respect to the chemical stability of the substance to be examined. So the reason for limiting its presence is in most cases that potassium is used in many raw material and reagents involved in chemical manufacturing, and a potassium residue, therefore, is a sign of insufficient purification. As with the other alkali metals, potassium is exclusively found as the monovalent cation. Since nearly all salts of potassium are water soluble it is a very inert ion which, as a main rule does not contribute to the characteristics of its salts. Ammonium is almost identical in size to potassium, and therefore has properties which are almost identical to this.

The test is based on the opalescence formed when potassium precipitates with the anionic reagent tetraphenylborate (Figure 6.11.1). This reagent has found widespread use in the determination of potassium by gravimetric, titrimetric, and as in the case of the pharmacopoeial limit test, turbidimetric analysis.

To 10 ml of the prescribed solution, 2 ml of a freshly prepared 10 g/l solution of *sodium tetraphenylborate R* is added. A standard is prepared in the

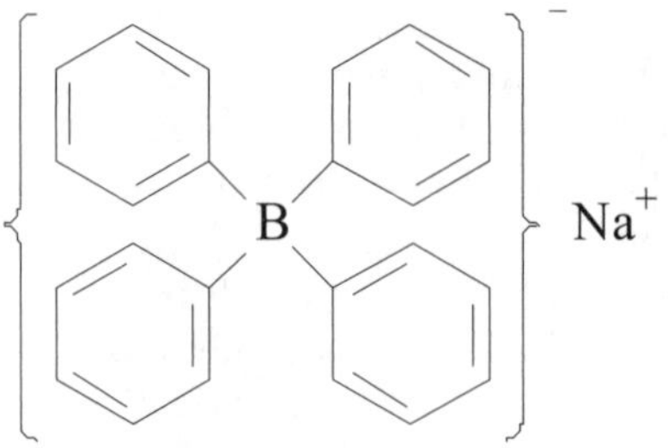

Figure 6.11.1 Sodium tetraphenylborate.

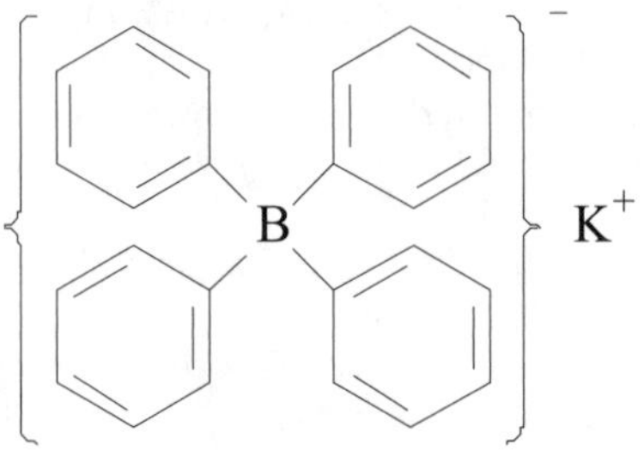

Figure 6.11.2 Potassium tetraphenylborate.

same manner using a mixture of 5 ml of *potassium standard solution (20 ppm K)*
R and 5 ml of *water R*. After 5 min, any opalescence in the test solution should
not be more intense than that in the standard.

Since potassium tetraphenylborate (Figure 6.11.2) has a very low solubility compared to that of sodium tetraphenylborate, it will give a white
precipitate under the conditions given in the test. The precipitate is formed
in neutral, slightly acidic, and slightly alkaline condition and in the presence
of diluted acetic acid. It is soluble in a strong acid and base through disso-
ciation but also soluble as an ion pair in some organic solvents as for example
acetone.

The other alkali metals do not give a precipitate even if present in a high
concentration compared to potassium, but ammonium will give an ammonium tetraphenylborate precipitate identical to potassium tetraphenylborate.
Of less importance is that a precipitate is also formed with rubidium, caesium,
thallium, silver, and some organic nitrogen substances.[1] Other references add
copper and mercury as being capable of giving a positive reaction with
tetraphenylborate.[2]

Alkali earth metals, zinc, cobalt, nickel, and manganese might interfere
but can, if their presence is expected, be masked by EDTA. Valence 3 cations
such as aluminum, chromium, and iron can also interfere unless masked
with sodium fluoride.[3]

References

1. Kolthoff, I.M. and Elving, P.J., *Treatise on Analytical Chemistry*, Part II, Vol. 1, Interscience Publishers, New York, 1961, pp. 359, 373, 439.
2. Kodama, K., *Methods of Quantitative Inorganic Analysis*, Interscience Publishers, New York, 1963, p. 409.
3. Fries, J. and Getrost, H., *Organische Reagenzien für die Spurenanalyse*, E. Merck, Darmstadt, 1975, p. 173.

6.12 Sulfates (European Pharmacopoeia 2.4.13)

The test ensures that the substance to be examined has a total content of sulfate (SO_4^{2-}), which is below the limit defined in the individual monograph. Sulfate is the most common of the many oxo acids of sulfur, and one of the most stable. It is a very water-soluble anion forming insoluble salts with only a few cations, among them lead, strontium, and barium. Sulfate is the corresponding base or acid radical of sulfuric acid, H_2SO_4, and since it is one of the cheapest of the strong mineral acids it is widely used in the synthesis and purification of chemicals. The reason for limiting its presence in active pharmaceutical ingredients and excipients is because in most cases it reveals an unsuccessful removal of the acid during purification. As a consequence of the widespread use of sulfuric acid, sulfate is referenced in several hundred monographs.

The sulfate content is estimated by precipitating the sulfate present in the substance to be examined with barium, and comparing the resulting opalescence to the opalescence obtained in a standard of known sulfate content. It is the same chemistry used in the pharmacopoeia identification test of sulfates.

$$Ba^{2+} + SO_4^{2-} \rightarrow BaSO_4\downarrow$$

Barium sulfate is a hard crystal, which has a solubility that varies dramatically with particle size. The characteristic behavior of such crystals in relation to precipitations like the present test is discussed in Chapter 4, "Precipitation in Limit Tests," in many cases using barium sulfate as an example.

It is stated in the start of the text that all solutions used should be prepared with *distilled water R*. The main reason for this is that the content of various ions could influence the degree of opalescence obtained in the test.

Step one in the test is to prepare a crystal seed by adding 1 ml of a 250 g/l solution of *barium chloride R* to 1.5 ml of *sulfate standard solution (10 ppm SO₄) R1*. Shake and allow to stand for 1 min. Then, 15 ml of the solution to be examined and 0.5 ml of *acetic acid R* are combined. Prepare a standard in the same manner using 15 ml of *sulfate standard solution (10 ppm SO₄) R1* instead of the solution to be examined. After 5 min, any opalescence in the test solution should not be more intense than that in the standard.

The precipitation in step one, where a small amount of sulfate is added in excess amount of barium reagents, is aimed at giving a precipitate of well-defined particle size distribution. The reagent concentration is high compared to the one seen when a sample is added and the *sulfate standard solution (10 ppm SO₄) R1* contains 30% (V/V) alcohol. The relative high concentration and the limit solubility of sulfate in alcohol mean that a high level of supersaturation is achieved, and this gives a fast precipitation with smaller particles. As also seen in 6.3. Calcium, the seeding step greatly enhances both sensitivity and reproducibility. Later when the sample solution is added the crucial step of nucleation is eliminated, and a further precipitation will occur immediately.

Experiments have revealed that the way the final test solution is handled is very important. Different results are obtained depending on how the 15 ml sample solution and the 0.5 ml acetic acid are added, for example whether reagents are thrown together or mixed gently.[1] Contrary to, for example, the limit test for calcium, precipitating sulfate with barium showed no sensitivity toward the temperature at which the procedure is performed. In conclusion, to obtain a reproducible and accurate result one must exercise the greatest care in performing the procedure, and the limited reproducibility of the test should be remembered in case of unexpected results.

The selectivity of the test is fairly good since the other common acid radicals that form white precipitates with barium in neutral solutions (carbonate, sulfite, and phosphate) fail to do so in dilute acetic acid.

The presence of farren salts interferes with the test result, and especially nitrate and oxalate lower the sensitivity of the test. When present they apparently inhibit the nucleation process, giving a slower onset of the precipitation, and in addition they lower the maximum opalescence reached in the test.[2]

A very through investigation into the effect of operational parameters on the result obtained in the limit tests of calcium and sulfates has been published by Zimmermann et al.[3]

References

1. Thörn, N., Kvantitativ prövning för påvisande av sulfat, *Farmacevtisk revy,* 41, 1942, 445.
2. Reimers, F., *The Basic Principles for Pharmacopoeial Tests*, Heinemann Medical Books, London, 1956, p. 57.
3. Zimmermann, J., Krogh-Svendsen, E., and Reimers, F., Limit tests for impurities IV. Investigations into the reproducibility of precipitation part III, *Analytica Chim. Acta*, 16, 1957, 6.

6.13   Sulfated ash (European Pharmacopoeia 2.4.14)

The test ensures that the substance to be examined has a content of inorganic cations capable of forming nonvolatile sulfate salts, which is below the limit defined in the individual monograph. These unintended cations could either be present as a consequence of contamination, but most often they will be residues of inorganic reagents and chemicals used in the synthesis of the substance to be examined. So the typical residue contained in the sulfate salt is the most widely used alkali metals, sodium and potassium. Since these cations are in most cases easily eliminated in the purification steps, the test gives a very important insight into the general quality in the production process of the substance to be examined, and it is therefore referenced in several hundred monographs. The test is predominantly used in testing chemically pure substances, whereas the limit test 6.14. Total ash is used for vegetable drugs. The sulfated ash is preferred to the total ash and other ashing procedures (see 6.14. Total ash) since it gives a more well-defined ash and thereby a more reproducible result. In a few monographs the test is used as a way to determine, for example, sodium deliberately present in the substance to be examined as a counter-ion.

Prior to handling the sample the crucible is heated to redness before determining the tare mass. This is crucial since a dirty crucible will give a meaningless result in the end.

A suitable crucible (silica, platinum, porcelain, or quartz) is ignited at $600 \pm 50°C$ for 30 min, allowed to cool in a desiccator over silica gel, and weighed. The prescribed amount of the substance to be examined is placed in the crucible and weighed. The substance to be examined is moistened with a small amount of *sulfuric acid R* (usually 1 ml) and heated gently at as low a temperature as practicable until the sample is thoroughly charred.

In this first heating the sample, being an organic compound, is charred to a charcoal-like substance. Since different compounds behave differently to this procedure, it should be carried out carefully at the lowest possible temperature, usually with a Bunsen burner. If excessive heat is used the sample could catch fire, or the char could burst or sputter. In both cases material will be lost. If the char does not sinter on the button of the crucible but stays as a ring on the walls of the crucible it will not be exposed to as much heat as if it had been on the bottom, and should be scraped down with a spatula or heated from the side with the burner. This is preferred to just raising the temperature, because of the risk of losing the residue through evaporation, especially if a platinum crucible is being used since it insulates very poorly. If it is heated above bright redness (700°C), materials like alkali metal chlorides are at risk of being lost. The procedure is carried out with the lid off, since the combustion process acquires oxygen.

After cooling, the residue is moistened with a small amount of *sulfuric acid R*, heated gently until white fumes are no longer evolved, and ignited at $600 \pm 50°C$ until the residue is completely incinerated. Ensure that flames are not produced at any time during the procedure. The crucible is allowed to cool in a desiccator over silica gel, it is weighed again, and the mass of the residue is calculate. If the mass of the residue so obtained exceeds the prescribed limit, it is moistened with *sulfuric acid R* and ignition is repeated, as previously, to constant mass, unless otherwise prescribed.

In this second step of analysis, the charred substance obtained by the first procedure is incinerated to evaporate all the carbon as carbon dioxide. This can be difficult to obtain if the substance to be examined was not fully charred before incineration. The residue obtained should be a well-defined and relatively stable mixture of the sulfate salts of the cations originally present, typically sodium sulfate and potassium sulfate. This gives a more reproducible result than if the test were based on a residue of the salts originally present since many of them are not stable enough to survive the ashing procedure. This applies, for example, to many carbonates, but even sulfate salts can be degraded if prolonged and excessive heat is applied.

It is also worth noticing that weighing to constant mass is not a requirement, if the result obtained is below the prescribed limit. This is new in the revised sulfated ash that became official in *European Pharmacopoeia* IV, which is a result of the international pharmacopoeia harmonization effort. Another difference in the new test is that the old test included a step where sodium carbonate was added to the residue with the purpose of eliminating certain pyrosulfates that allegedly could build up as a result of igniting with sulfuric acid.

Since the basis of the test is to weigh the residue remaining in the substance to be examined after ignition, the detection limit obtainable is restricted, in principle, by the amount of sample brought to work and the accuracy of the mass determination. A limit below 0.1% would therefore require a sample of more than 1 g if no higher uncertainty on the mass determination were allowed. Therefore, if the presence of some of the toxic cations, as for example lead, is suspected in the substance to be examined, a special test has to be applied as they have limits in the area of 20 ppm.

Since the procedure depends on determining a small difference in the mass of a crucible before and after it is exposed to harsh conditions, great care should be exercised in carrying out the determination. Procedures should be in place to ensure that the crucible is as clean a possible, even before the initial cleaning incineration is carried out. If the crucible is contaminated despite this incineration, chances are the contaminant will evaporate during the, sometimes prolonged, sample incineration, giving a negative residue mass. Since small differences in mass have to be determined, procedures regarding the weighing have to be standardized. The desiccator step can cause problems when a difference in relative humidity, degree of filling, and ambient temperature affords that the crucible temperature is different in the weightings before and after sample incineration. If the crucible during weighing has a temperature different from the temperature in the weighing cabinet, the deviation between the density of the air inside the crucible and outside the crucible will cause a bias. Static electricity is a potential problem, especially in the dry air of a heated laboratory during cold seasons. Often it is advisable to use the same weight throughout the procedure. Finally, it should be remembered that most crucibles, except platinum, will lose some weight during the procedure, especially old ones.

In general, this is a test that puts rather high demands on the skills and patience of the operator performing the analysis. The problems this entails can largely be eliminated by using an automatic wet digestion apparatus that can be programmed to perform either the initial charring or the whole procedure unattended and at the lowest possible temperature. Furthermore there is an environmental benefit since the apparatus scrubs the sulfuric acid fumes evolved, eliminating emission.[1–3]

References

1. Reimers, F., *The Basic Principles for Pharmacopoeial Tests*, Heinemann Medical Books, London, 1956, pp. 40, 41.
2. Reimers, F. and Gottlieb, K.R., *Contributions from the Danish Pharmacopoeia Commission*, Vol. 1, *Limit Tests for Impurities*, Ejnar munksgaard, København, 1946, pp. 136, 162.
3. Council of Europe, *Pharmeuropa*, March 1997, Council of Europe, Strasbourg, 1997, p. 127.

6.14   Total ash (European Pharmacopoeia 2.4.16)

The test ensures that the substance to be examined has a content of inorganic ash, which is below the limit defined in the individual monograph. Total ash is one of the three ash values, which traditionally have been used in the evaluation of vegetable drugs: total ash, acid insoluble ash, and sulfated ash. The total ash is the residue obtained when the vegetable drug is carefully ignited to burn off all the carbon. The residue consists of inorganic salts either naturally present in the vegetable drug or brought there by contamination. The acid soluble ash is the part of the total ash that is insoluble in dilute hydrochloric acid, and this is often an indication of the vegetable drug being contaminated with earthy material, for example sand. Some vegetable drugs can, however, have a very high natural content of acid insoluble ash in the form of silica. In the sulfated ash procedure sulfuric acid is added to the vegetable drug sample before and during the incineration. By this procedure most salts are converted to sulfate salts, which due to their low volatility lowers the risk of losing a part of the residue through evaporation. This greatly enhances the reproducibility of the determination. The sulfated ash procedure is widely used as a quality control parameter for active pharmaceutical ingredients and excipients as well. At the present a few hundred monographs reference total ash. Most of them describe vegetable drugs, ranging from whole dried leaves (e.g., Belladonna leaf), and powdered crude plant material (e.g., Ipecacuanha, prepared), to more chemically well-defined extracts (e.g., Carnauba wax). Limits defined in the individual monographs vary from 0.1% to 25%, but typically values are between 5% and 7%. Pure chemicals are usually tested using 6.13. Sulfated ash.

A silica or platinum crucible is heated to redness for 30 min, allowed to cool in a desiccator, and weighed. Unless otherwise prescribed, 1 g of the

substance or the powdered vegetable drug to be examined is evenly distributed in the crucible. It is dried at 100°C to 105°C for 1 h and ignited to
constant mass in a muffle furnace at 600°C ± 25°C, allowing the crucible to
cool in a desiccator after each ignition. Flames should not be produced at
any time during the procedure. If after prolonged ignition the ash still contains black particles, it is taken up with hot water, filtered through an ash-less
filter paper, and the residue and the filter paper is ignited. The filtrate and
the ash are combined, carefully evaporated to dryness, and ignited to constant mass.

In the first step of analysis, the crucible is heated to redness before
determining the tarra mass. This step is crucial since a dirty crucible will
give a meaningless result in the end. Taking a 1 g sample of a vegetable drug
can, especially in the cases of crude plant material, constitute a problem.
Homogeneity can be ensured by grinding a larger portion. The drying procedure is necessary to avoid the risk of losing sample material due to bursts
of evaporation when the water content is exposed to strong heat in the muffle
furnace. The risk of bursts, however, is not completely eliminated through
drying since a content of resin can also make it sputter. Ashing methods
described in earlier pharmacopoeias often start with a careful treatment over
a Bunsen burner, so that elimination of water and material decomposition
without flames could easily be monitored.

The typical residue obtained, disregarding mineral contamination, consists of sodium, potassium, and calcium as their carbonate, chloride, sulfate,
and phosphate salts, in addition to silicates. Ammonium salts are evaporated
during the procedure, and carbonates and even sulfates can be lost at temperatures higher than the defined maximum of 625°C. Even alkali salts can
be lost if excessive heat is used. Mixtures of sodium chloride and potassium
chloride melt at approximately 660°C, so in general the lowest possible
temperature should be used.

If the vegetable drug has a high content of alkali salts, it can hinder the
combustion, leaving black particles in the residue. This is fixed by taking it
up with hot water.

Since the procedure depends on determining a small difference in the
mass of a crucible before and after it is exposed to harsh conditions, great
care should be exercised carrying out the determination. Procedures should
be in place to ensure that the crucible is as clean as possible, even before the
initial cleaning incineration is carried out. If the crucible is contaminated,
despite this incineration, chances are the contaminant will evaporate during
the, sometimes prolonged, sample incineration, giving a negative residue
mass. Since small differences in mass have to be determined, procedures
regarding the weighing have to be standardized. The desiccator step can
cause problems when a difference in relative humidity, degree of filling, and
ambient temperature affords that the crucible temperature is different in the
weightings before and after sample incineration. If the crucible during
weighing has a temperature different from the temperature in the weighing

cabinet, the deviation between the density of the air inside the crucible and outside the crucible will cause a bias. Static electricity is a potential problem, especially in the dry air of a heated laboratory during cold seasons. Often it is advisable to use the same weight throughout the procedure. Finally it should be remembered that most crucibles, except platinum, will lose some weight during the procedure, especially older ones.[1,2]

References

1. Evans, W.C., *Trease and Evans' Pharmacognosy*, 13th ed., Bailloère Tindall, London, 1989, p. 131.
2. Hänsel, R., Sticher, O., and Steinegger, E., *Pharmakognosie — phytopharmazie*, 6. auflage, Springer-Verlag, Berlin, 1999, p. 139.

6.15   Free formaldehyde (European Pharmacopoeia 2.4.18)

The test ensures that the substance to be examined has a total content of formaldehyde (H_2CO), which is below the limit defined in the individual monograph. The monograph is named free formaldehyde to exclude its use on formaldehyde groups chemically bound to the substance to be examined. The test is referenced in about 30 monographs all describing vaccines, all of them referencing method (a) and nearly all of them having a free formaldehyde limit of 0.2 g/l. Vaccines are inactivated virus particles or virus infected cells. Both contain specific antigens, very often glycoproteins, which in the organism given the vaccine raises an immune response. Formaldehyde is used to kill the virus or virus host without destroying the vaccine glycoproteins. Since residual free formaldehyde, due to its toxicity, is unwanted in the final vaccine, it is removed by a number of techniques. One of these is neutralization, using sodium metabisulfite through the reaction shown below. It reacts very readily with many substances and polymerizes easily.

$$CH_2O + NaHSO_3 \rightarrow CH_2(OH)SO_3Na$$

The *European Pharmacopoeia* has two different colorimetric methods for limit testing of free formaldehydes based on a condensation of the formaldehyde being determined with acetylacetone and MBTH respectively.

Method (a)

This method first described by Nash in 1952 has the advantage over many of the other colorimetric formaldehyde methods in that the reaction takes place under relatively mild reaction conditions. One disadvantage, however, is that a reaction is gained not only with free formaldehyde but to some extent also with bound and, although to a limited extent, polymeric formaldehyde.

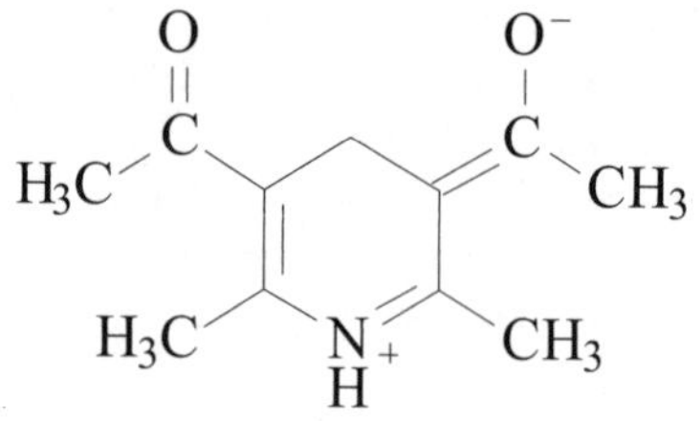

Figure 6.15.1 3,5-diacetyl-1,4-dihydrolutidine.

For vaccines for human use, a 1 in 10 dilution of the vaccine to be examined is prepared. For bacterial toxoids for veterinary use, a 1 in 25 dilution of the vaccine to be examined is prepared.

To 1 ml of the dilution, 4 ml of *water R*, and 5 ml of *acetylacetone reagent R1* are added. The tube is placed in a water-bath at 40°C for 40 min and afterward examined down their vertical axes. The solution is not more intensely colored than a standard, prepared at the same time and in the same manner, using 1 ml of a dilution of *formaldehyde solution R* containing 20 μg of formaldehyde (CH_2O) p/ml, instead of the dilution of the vaccine to be examined.

Under these conditions formaldehyde, acetylacetone, and ammonia react to give a yellow product called 3,5-diacetyl-1,4-dihydrolutidine (Figure 6.15.1).

This product, of which only one mesomeric form is shown, has an absorbance that obeys Beer's law. When formaldehyde concentration is low, as in the present test, reaction is claimed to be nearly quantitative in the pH range 5.5 to 6.5. Apparently it follows first order kinetics and also follows the Arrhenius relation closely. This imparts that a 99% completion of the reaction requires 40 min at 37°C and 5 min at 58°C. It is fairly insensitive toward variations in reagent concentration, but least insensitive with regard to ammonia. It should be remembered that all reagents including formaldehyde are volatile. It fades slowly in the order of a few percentages overnight but more rapidly (25%) when submitted to diffuse daylight all day. The degradation is slowed if solutions are kept in stoppered test tubes. The conjugated product is soluble in polar solvents but insoluble in ether, and the intensity of its color is insensitive toward pH changes in the range of 4 to 10.

Acetic aldehyde also gives a similar product but only in a yield of about one percentage if present in the same amount as formaldehyde. Acetone, chlorale, furale, glucose, glycerine, and ninhydrine does not interfere. Some amines, like methylamine and ethylenediammine, can compete with ammonia in the reaction causing less product to be formed. As already stated polymeric formaldehyde does react to some extent, most likely through release of free formaldehyde, but is reacts slowly. Sulfite prevents the reaction almost completely at 0.001 sodium sulfite. Oxidizing agents will destroy the colored product.

It has been shown that the method works fine for vaccines not neutralized, but gives high results when neutralized with sodium metabisulfite,
apparently because it reacts with the Nash reagent.[1,2]

Method (b)

Test solution (a) in 200 dilution of the vaccine to be examined with R is
prepared as described in the pharmacopoea. Solutions containing 0.25 g/l,
0.50 g/l, 1.00 g/l, and 2.00 g/l of CH_2O by dilution of *formaldehyde solution
R* with *R* are prepared. To 0.5 ml of the test solution and of each of the
reference solutions in test tubes, 5.0 ml of a freshly prepared 0.5 g/l solution
of *methylbenzoethiazolone hydrazone hydrochloride R* is added. Tubes are closed,
shaken, and allowed to stand for 60 min. One ml of *ferric chloride-sulfamic
acid reagent R* is added and allowed to stand for 15 min. The absorbance of
the solutions is measured at 628 nm, and the content of formaldehyde in the
vaccine to be examined is calculated from the calibration curve established
using the reference solutions. The test is invalid if the correlation coefficient
(r) of the calibration curve is less than 0.97.

It is explicitly stated in the pharmacopoea that the method is reserved
for vaccines where the excess formaldehyde is neutralized with sodium
metabisulfite.

It is a color reaction based on the reagent 3-methylbenzothiazolin-2,1-
hydrazone, most often abbreviated to MBTH (Figure 6.15.2). The reagent has
been widely used for determining and identifying aldehydes and a variety
of other substances, and several articles are quoted in the book referenced
below.

Under the conditions given two molecules of MBTH will condense with
one molecule of formaldehyde, yielding water and the product given in
reaction (Figure 6.15.3). This product is oxidized by the iron(III) present
giving a molecule of high conjugation.

The test is described as the most sensitive of the contemporary colorimetric methods. The absorbance follows Beer's law and has a maximum at
635 nm. The position of the maximum varies, however, with the solvent
used. The sulfamic acid, which is added to the reaction via the ferric chloride
reagent, ensures that the colored compound does not precipitate, leaving the
test solution turbid. This was often seen in older variants of the method, and

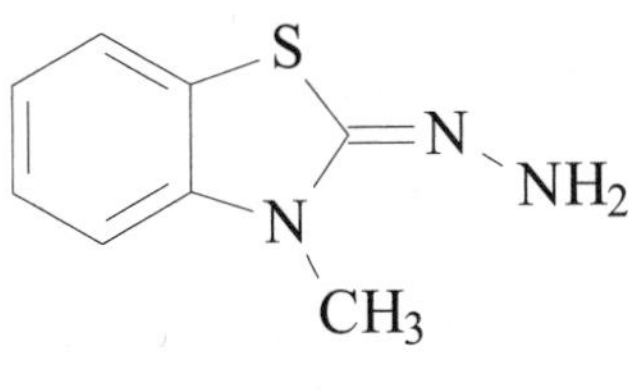

Figure 6.15.2 3-methylbenzothiazolin-2,1-hydrazone.

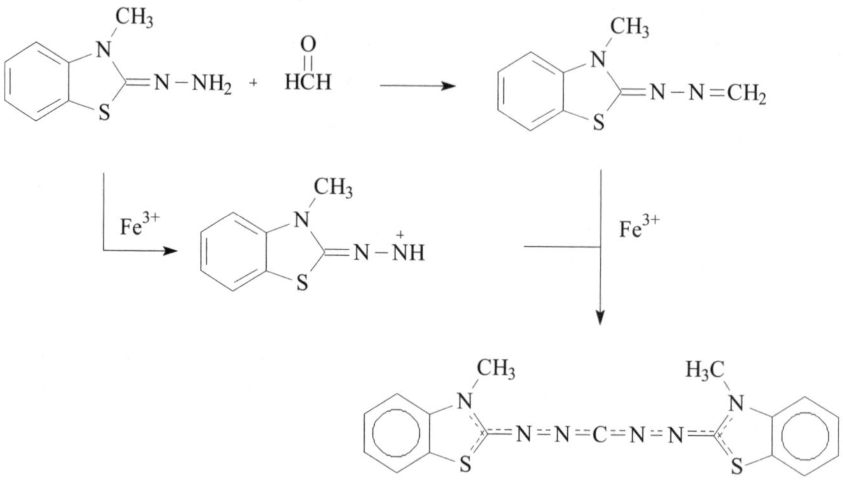

Figure 6.15.3 Reaction.

Table 6.8 Wavelength Maximum and Absorbance of Different Aldehydes

	Solvent	λ Max nm	A = 0.3 Sample µg
Formaldehyde	Water	635	0.87
Acetaldehyde	Water	610	1.25
Propionaldehyde	Water	620	1.68
Butyraldehyde	Water	615	2.0
Caprylic aldehyde	Isopropanol	635	4.6
Lauric aldehyde	Isopropanol	635	33.6
Citronellal	Isopropanol	640	11.6
Citral	Isopropanol	640	13.5

it enforced that extraction into acetone was needed.[3] The reagent has, as earlier mentioned, been used for many different substances in the past. When the exact procedure is applied to other aliphatic aldehydes, a condensation product analogue to the above shown is formed, but the position of the wavelength maximum and the intensity differs from the one seen with formaldehyde as shown in Table 6.8.

Even aromatic aldehydes give a reaction, but Beer's law is not always followed. When other techniques are used the reagent can be used for determining primary alcohols, aromatic aldehydes, ethylenic compounds, aldoses and ketoses, 2-amino-2 deoxyhexoses, and 17-hydroxy-17-ketolsteroids.[4]

References

1. Taylor, E.M. and Moloney, P.J., Estimation of aldehyde in poliomyelitis vaccines, *J. Amer. Pharm. Assoc.*, 46, 299, 1957.
2. Nash, T., The colorimetric estimation of formaldehyde by means of the hantzscch reaction, *Biochem. J.*, 55, 416, 1953.

3. Kakáč, B. and Vejdělek, Z.J., *Handbuch der photometrishe analyse organisher verbindungen*, Band 1, Verlag Chemie, Weinheim, 1974, p. 232.
4. Pesez, M. and Bartos, J., *Colometric and Fluorimetric Analysis of Organic Compounds and Drugs*, Marcel Dekker, New York, 1974, pp. 264, 536.

6.16   Alkaline impurities in fatty oils (European Pharmacopoeia 2.4.19)

This test ensures that the fatty oil to be examined has a content of alkaline impurities, which is below the limit defined in the individual monograph. The need for controlling alkaline residues in fatty oils comes from one of the procedures used to purify crude vegetable oil into pharmaceutical grade oil.

The fatty oils covered by the purpose of the test are triglycerides of miscellaneous fatty acids. Triglycerides, called fixed fats or fixed oils, are used for the storage of energy in both vegetable and animal cells and can be isolated from these sources by extraction with lipofilic solvents, by centrifuging, or by pressing. The risk of transferring viruses from raw material of animal sources has reduced the use of oils of animal origin, but formerly, for example, whale fat was used directly.

A triglyceride is the glycerine ester of different monocarboxylic fatty acids, and the actual fatty acid decides the properties of the oil. Saturated fatty acids produce an oil that becomes solid at relatively high temperatures, and these are normally referred to as fats. Unsaturated fatty acids solidify normally at temperature below ambient and are referred to as oils. Naturally occurring oil most often has different fatty acids on the three alcohol groups, and the oil from different sources has a content of fatty acids that is characteristic for the source. Below is a schematic triglycerid of soja oil, refined (Figure 6.16.1). The fatty acids are linoleic acid, which according to the demands of European *Pharmacopoeia* must constitute 48 to 58% of the total fatty acid content (on top), oleic acid in 17 to 30%, and palmitic acid in 9 to 13% (lower).

The most common vegetable sources of oils are fruits and nuts, and the crude oil obtained by either processed will contain a variety of unwanted lipofilic constituents that have to be removed and altered by the processes of refining, bleaching, hydrogenation, and deodorizing. In the process of refining, free fatty acids are precipitated as insoluble alkali metal salts, usually by the extraction with an alkali hydroxide solution. The free fatty acid stays

Figure 6.16.1 Schematic presentation of soja oil fatty acid.

in the water as soap, but a small fraction might stay in the oil as an alkaline impurity. Such a residue reduces the stability of the oil. The purpose is therefore to test the potential residue of alkaline equivalents unintentionally present in the substance to be examined.[1] A more detailed explanation of the oil refining process can be found in the general monograph Vegetable fatty oils in the *European Pharmacopoeia*. At present only two monographs reference the test: Coconut oil, refined and Soya-bean, hydrogenated. Coconut oil has another limit than the general one, and there is a full description of the test in that monograph.

In a test tube, mix 10 ml of recently distilled *acetone R* and 0.3 ml of *water R* and add 0.05 ml of a 0.4 g/l solution of *bromophenol blue R* in *alcohol R*. Neutralize the solution if necessary with *0.01 M hydrochloric acid* or *0.01 M sodium hydroxide*.

Add 10 ml of the oil to be examined, shake, and allow to stand. Not more than 0.1 ml of *0.01 M hydrochloric acid* is required to change the color of the upper layer to yellow.

Any free fatty acid present in the substance to be examined, for example a sodium salt, will, by the procedure, be dissolved in acetone and hereafter titrate by hydrochloric acid (see equation below). The content of alkaline impurities is at or below the defined limit, and the excess of hydrochloric acid added will change the color of the indicator from blue to yellow. Of course any other alkaline substances present capable of neutralizing the added hydrochloric acid will act as free fatty acid bases.

$$CH_3[CH_2]_{14}COONa \quad + \quad H^+ \quad \longrightarrow \quad CH_3[CH_2]_{14}COOH \quad + \quad Na^+$$

Reference

1. Hänsel, R., Sticher, O., and Steinegger, E., *Pharmakognosie — phytopharmazie*, 6. auflage, Springer-Verlag, Berlin, 1999, p. 209.

Index